Environmental Toxicology

MAN AND THE ENVIRONMENT INFORMATION GUIDE SERIES

Series Editor: Seymour M. Gold, Associate Professor of Environmental Planning, Department of Environmental Planning and Management, University of California, Davis, California

Also in this series:

AIR POLLUTION—*Edited by George Hagevik**

ENERGY PLANNING—*Edited by David Myhra**

ENVIRONMENTAL ECONOMICS—*Edited by Cleve E. Willis and Barry C. Field**

ENVIRONMENTAL EDUCATION—*Edited by William B. Stapp and Mary Dawn Liston*

ENVIRONMENTAL LAW—*Edited by Mortimer D. Schwartz*

ENVIRONMENTAL PLANNING—*Edited by Michael J. Meshenberg*

ENVIRONMENTAL VALUES, 1860-1972—*Edited by Loren C. Owings*

NOISE POLLUTION—*Edited by Clifford R. Bragdon**

WASTE DISPOSAL IN MARINE ENVIRONMENTS—*Edited by Harold D. Palmer**

WASTEWATER MANAGEMENT—*Edited by George Tchobanoglous, Robert Smith, and Ronald Crites*

WATER POLLUTION—*Edited by Allen W. Knight and Mary Ann Simmons**

*in preparation

The above series is part of the

GALE INFORMATION GUIDE LIBRARY

The Library consists of a number of separate series of guides covering major areas in the social sciences, humanities, and current affairs.

General Editor: Paul Wasserman, Professor and former Dean, School of Library and Information Services, University of Maryland

Managing Editor: Denise Allard Adzigian, Gale Research Company

Environmental Toxicology

A GUIDE TO INFORMATION SOURCES

Volume 7 in the Man and the Environment Information Guide Series

Robert L. Rudd

Professor of Zoology
University of California, Davis

Gale Research Company
Book Tower, Detroit, Michigan 48226

Library of Congress Cataloging in Publication Data

Rudd, Robert L
Environmental toxicology.

(Man and the environment information guide series; 7)
(Gale information guide library)
Includes indexes.
1. Pollution--Environmental aspects.
2. Pesticides--Environmental aspects. I. Title.
QH545.A1R8 301.31 73-17540
ISBN 0-8103-1342-1

VITA

Robert L. Rudd is a professor of zoology at the University of California, Davis. Born in Los Angeles in 1921, he was reared in Salinas, California, and educated at the University of California, Berkeley. His doctorate was granted in zoology from Berkeley in 1953. Rudd enlisted in the Army Air Corps in 1942 and for three years was a combat pilot. This military experience initiated his first foreign travel, in this instance to Australia, New Guinea, Indonesia, and the Philippines. Since that time he has traveled to many countries, attending many conferences and lecturing at several universities.

Basically his interests are in ecology and evolution, but in 1952 he began pursuing his interest in the impact of man-induced chemicals on wildlife and ecosystems. This interest has expanded to include the entire range of chemical effects on matters of importance to man and the social consequences of technological invention. He has authored some sixty titles, the most notable of which is PESTICIDES AND THE LIVING LANDSCAPE (1964). Rudd continues to work on ecosystems "disturbed" by man and on correctives that need to be installed for man's benefit. He serves as advisor to many organizations and governments.

CONTENTS

Contents

FOREWORD

The rapidly expanding scope and sophistication of literature in the area of environmental studies has created the need for an authoritative, annotated guide to the range and quality of information sources. New legislation, publications, developing techniques, and increasing levels of public involvement in environmental issues have created a need for government, industry, organizations, and individuals to seek more authoritative information on man-environment relationships. This information can be used to rationalize proposed public or private actions, resolve issues, and measure the impact of past actions on man and the environment.

The objective of this series is to provide an authoritative and systematic guide to significant information sources on selected topics for use by scholars, students, scientists, reference librarians, consultants, professionals, and citizens. The scope and scale of each volume are directed toward accommodating both the immediate and continuing needs of users in addition to establishing a chronological benchmark of the literature on each topic.

This series is concerned with the cause and effect relationships of man's impact on the urban and natural environment. It emphasizes the human problems, issues, and implications of these relationships that can be used to study or solve environmental problems. The series is intended to be a primary reference for people not familiar with the location and organization of information sources in the general area of environmental studies. It should also serve as a basic reference for advanced students or professionals who need an annotated and selected sample of the current literature.

Each volume is prepared by a noted authority on the topic. The annotations and selection of references have been carefully reviewed to present an objective and balanced array of thought and practice. The organization of each volume is tailored to emphasize the way information is organized in this multidisciplinary and rapidly evolving field.

This volume is the most authoritative, current, and comprehensive reference work on environmental toxicology available. It brings together a diverse set of information sources from the physical, social, and natural sciences.

Foreword

The volume presents a compendium of reference material and ideas invaluable to the scholar and practitioner. It contains a highly selected collection of periodical literature, government documents, scientific journals, and teaching materials on the rapidly evolving field of environmental toxicology. The introductions, annotations, and sources have been carefully written to give the reader a technical and pragmatic grasp of problems involved in the scientific and applied aspects of environmental toxicology.

The scholarly nature of annotations, classification, and cross-referencing of materials presents an integrated view of time, space, values, and events that is not common to other reference works in this area. The author has made a very technical and unorganized subject accessible to practitioners and students by his clear and concise organization of a wide range of sources.

This is a major contribution by a noted authority on the literature of the field. It should help to improve the science of environmental toxicology. Environmental toxicology has become a critical dimension of living in a world of limited resources. This volume contains a blend of technical information from professional practice and scholarly thought that can help to create and preserve a quality of life and environment still possible on Spaceship Earth.

As the seventh volume in this series, ENVIRONMENTAL TOXICOLOGY continues the standard of scholarly excellence and problem solving usefulness that will be followed in additional volumes now under contract. When the entire series is completed in 1979, it will represent the most authoritative and comprehensive guide to the man-environment literature ever published.

Seymour M. Gold, Series Editor
August 1977

INTRODUCTION

At first glance the term environmental toxicology would seem to be easily understandable. It is the study of the actions of toxic substances in any aspect of environment. But a simple definition is not very helpful. There are many environments, thousands of toxic substances, a seeming infinity of interactions, and a welter of information from dozens of scientific disciplines. Rather than become hopelessly overwhelmed with the recognition that no one person can possibly understand all technical and interpretative information in this field, let us take a positive stance and state what we can do and learn.

The first thing to recognize is that, whatever the precise aspect, environmental toxicology is a highly technical subject. Intensive training is required in all the disciplines included within its rubric. Specialists may approach the subject from the ultimate technical frontiers of their disciplines and look upward toward a group of integrating concepts among disciplines. Or generalists might look downward from these concepts and carry their analyses as far into a discipline as needed to establish the validity of an idea. More commonly, scientists adopt a mixture of the two approaches, adding particular competence in one or more disciplines to lesser knowledge of others. Judicious self-education, consultation, and opinion lead to reliable general conclusions.

Judgments in environmental toxicology must be synthesized from disparate data. That is the key to the study of this important field. The general and technical references listed in this source book have been assembled with synthesis in mind.

There are, of course, immediate and narrower questions to be answered. One might want to know what the safe level of DDT is to fishes or to man. Or, what are the environmental conditions which contribute to the conversion of inorganic mercury to its biologically hazardous methylated form. Hundreds of questions such as these can be answered by the information in this book.

More importantly, the sources given here have in turn their own reference lists. Recent publications were deliberately chosen; or, older publications were chosen as the best possible reference. These citations are the gateway to an immense literature. I estimate that the some 1,000 citations in this book have at least 125,000 references listed in them. The information on environmental toxicology is truly staggering.

Literature appears at an increasing rate. Problems with toxic substances are by no means diminishing with increasing regulation. Industrial civilization and social complexity intensify and extend the problems. Greater demands for knowledge, judicious interpretation, and social applications in environmental toxicology can be confidently predicted.

Nor can one consider the technical side of this field as the jurisdiction of research scholars and scientists only. Environmental toxicology exists because practical problems of great personal, economic, social, and legislative importance exist. We must constantly adapt to changing knowledge and demand. Our personal and social welfare depends on how well we make the adaptations.

These practical considerations are often surrounded by contention. Solutions are often compromises. The values placed on environmental utilization greatly differ according to the interests of the viewer. Frequently, in controversy, charges are made as a result of subjective or emotional presentation of views. I defend such elements of viewpoint as long as the best information is sought, the best argument for a set of values is mustered, and the majority interests of the society are served. We are, after all, people serving ourselves.

My own approach to environmental toxicology stems from my training as an ecologist and zoologist. The selection of materials reflects this background, with intentional stress on the environmental aspects of the subject. Limitations of space and emphasis require restricted representation of citations in chemistry, medicine, and pharmacology, as well as productive agriculture and forestry in which calculated pollution is a necessity. Moreover, several volumes in the Man and the Enviroment series address themselves to environmental pollution and its educational and legislative aspects. These volumes are companions in information and decision and should be consulted along with this volume.

A reader depends on sources. So also does an author and compiler. I have many individuals and organizations to thank in preparing this volume. My thanks are extended first to Betty Rudd for a long history of assistance and sufferance. I appreciate very much the editorial and typing assistance of Mrs. Mimi Williams. Finally, I owe much to my colleagues at the University of California, Davis. The overseeing guidance of Professor S.M. Gold, editor of the Man and the Environment series, has been most helpful. Most of all, without the unique library of the Food Protection and Toxicology Center on my campus, my task would have been much more difficult. I thank Dr. Ming-Yu Li and Judith Dohn who have made that library what it is.

Robert L. Rudd
University of California, Davis

Part I

GENERAL SOURCES ON ENVIRONMENTAL TOXICOLOGY

Section 1

ABSTRACTING SERVICES AND BIBLIOGRAPHIES

The literature from which environmental toxicology draws its information is immense. One must use data and ideas from many sources. Environmental toxicology is not a carefully defined scientific discipline. As in all studies of ecology and environment, one must accumulate and synthesize information from sources in widely disparate disciplines. Three elements are therefore present in judgments in environmental toxicology: immensity of data; diversity of data; and the necessity to integrate and to synthesize information. Sections 1 and 2 identify the diverse range of sources. Section 1 concerns organized retrieval systems without which integration would be next to impossible. Also valuable is the abundant and varied referencing in ENVIRONMENTAL EDUCATION, the first volume in the Man and the Environment series. Many of the journals, societies, and information sources described in that volume also apply to the subject of environmental toxicology.

1 ABRIDGED INDEX MEDICUS. Bethesda, Md.: National Library of Medicine, Bibliographic Services Division, 1970--. Quarterly.

Classified record of current world medical literature.

2 ABSTRACTS ON HEALTH EFFECTS OF ENVIRONMENTAL POLLUTANTS. Philadelphia: Biosciences Information Service (BIOSIS), 1972--. Monthly.

Index to publications; cross-referenced with short description of contents. Concentrates on industrial and occupational medicine.

3 AGRICULTURAL AND VETERINARY CHEMICALS. Seven Oaks, Kent, Engl.: Chandler Publishers, 1970--. Bimonthly. Approx. 350 pages per year.

Classified information on the manufacture, application, and current research in these classes of chemicals.

4 AIR POLLUTION ABSTRACTS. Research Triangle Park, N.C.: Environmental Protection Agency, Technical Information Center, 1970--. Monthly.

Cumulative author and subject indexes.

5 AIR POLLUTION TITLES. University Park: Center for Air Environment Studies, Pennsylvania State University, 1965--. Bimonthly. Approx. 500 pages per year.

Quick and efficient guide to current literature but has good capacity for retrospective search as well.

6 AMERICAN DOCUMENTATION. Washington, D.C.: American Society for Information Science, 1950--. Bimonthly. Approx. 500 pages per year.

Journal of the American Society for Information Science (ASIS); deals with storage and retrieval systems in the sciences particularly.

7 BIBLIOGRAPHY OF AGRICULTURE. New York: Macmillan Publishing Co., 1942--. Monthly. Approx. 4,800 pages per year.

Extremely useful index to the literature of agriculture and allied sciences; based on materials received by the National Agricultural Library.

8 BIOLOGICAL ABSTRACTS. Philadelphia: Bioscience Information Service of Biological Abstracts (BIOSIS), 1926--. Twice monthly. 2 vols. annually; not paginated. Approx. 70,000 numbered entries annually.

Standard and essential abstracting service; author and subject index; annotations and cross-referencing. Subject fields.

9 CANCER CHEMOTHERAPY ABSTRACTS. Bethesda, Md.: National Cancer Institute (National Institutes of Health), Program Analysis Branch, Division of Cancer Treatment, 1960--. Monthly. Not paginated. Approx. 5,000 entries per year.

10 CARCINOGENESIS ABSTRACTS. Bethesda, Md: National Cancer Institute (National Institutes of Health), Division of Cancer Cause and Prevention, 1962--. Monthly. Not paginated. Approx. 7,200 entries per year.

Subject divisions.

11 CHEMICAL ABSTRACTS. Columbus, Ohio: American Chemical Society, Chemical Abstracts Service, 1907--. Twice monthly; each issue paginated. Over 210,000 entries per year.

Comprehensive, worldwide listings presented in English. Essential indexing source.

12 CURRENT CONTENTS: AGRICULTURE, BIOLOGY AND ENVIRONMENTAL SCIENCES. Philadelphia: Institute for Scientific Information,

1970--. Weekly. Approx. 100 pages per issue.

Reports of current research in agriculture and biological, food, and veterinary sciences.

13 ENVIRONMENTAL ABSTRACTS. New York: Environmental Information Center, Publishing Division, 1971--. Monthly, except bimonthly summer. Approx. 1,800 pages per year.

Known formerly as ENVIRONMENTAL INFORMATION ACCESS.

14 ENVIRONMENTAL INDEX. New York: Environmental Information Center, 1971--. Annual. Approx. 800 pages per year.

Current aspects of environment in all phases. Includes conferences, activities, articles. Nontechnical subheadings.

15 ENVIRONMENTAL PERIODICALS BIBLIOGRAPHY. Santa Barbara, Calif.: Environmental Studies Institute, 1972--. Bimonthly. Approx. 700 pages per year.

Treats contamination of food, air, land, and water. Emphasizes resources.

16 EXCERPTA MEDICA. Amsterdam: Excerpta Medica, 1947--. Monthly.

Lists of articles from medical journals of all countries. Titles are usually given in English, along with original language. Massive journal. Forty-nine sections from anatomy to virology. Subject divisions.

17 FOOD SCIENCE AND TECHNOLOGY ABSTRACTS. Farnham Royal, Engl.: Commonwealth Agricultural Bureaux, 1969--. Monthly.

Monitors food sciences field.

18 HEALTH ASPECTS OF PESTICIDES ABSTRACT BULLETIN. Washington, D.C.: Environmental Protection Agency, Technical Services Division, Office of Pesticide Programs, 1968--. Monthly. About 3,200 abstracts per year; 1,000 pages per year.

Current summations of worldwide literature on pesticides and their effects.

19 INFORMATION STORAGE AND RETRIEVAL. New York: Pergamon Press, 1965--. Quarterly. Approx. 415 pages per year.

Medium for publication on theory and techniques of data storage and organization; special consideration to scientific information.

20 MICROBIOLOGY ABSTRACTS. London: Information Retrieval, 1965--. Monthly. Approx. 2,400 pages per year.

Covers a wide range: biochemistry, disease, toxins, plants, animals, man.

21 PESTICIDES DOCUMENTATION BULLETIN. Silver Spring, Md.: Network Information Services, 1965--. Monthly.

Chemical usage relating to insects, disease, nematodes, parasites, and other plant and animal relationships.

22 POLLUTION ABSTRACTS. Louisville, Ky.: Pollution Abstracts, 1970--. Bimonthly. About 6,000 citations per year; 1,000 pages per year.

Subject divisions. Indexes and summarizes world literature on environmental pollution--soil, air, and water. Covers pesticides and solid wastes. Essential index service.

23 REVIEW OF APPLIED ENTOMOLOGY. Series A, Agriculture; Series B, Medical and Veterinary. London: Commonwealth Agricultural Bureaux, Commonwealth Institute of Entomology, 1913--. Monthly. Approx. 1,000 pages per year.

Abstracts and reviews; worldwide coverage. Cumulative author index in each volume.

24 SELECTED REFERENCES ON ENVIRONMENTAL QUALITY AS IT RELATES TO HEALTH. Bethesda, Md.: National Library of Medicine, Bibliographic Services Division, 1971--. Monthly. Approx. 425 pages per year. Subject and author indexes.

25 SELECTED WATER RESOURCES ABSTRACTS. Washington, D.C.: Department of Commerce, Bureau of Standards, 1968--. Monthly. Approx. 2,400 pages per year.

Scientific and technological references on water conservation, quality, and management.

26 TOXICITY BIBLIOGRAPHY. Washington, D.C.: Government Printing Office (for the National Library of Medicine), 1968--. Quarterly. Not paginated.

Covers reports on toxicity studies, adverse drug reactions, and poisoning in man and animals. Subject divisions.

27 WATER POLLUTION ABSTRACTS. London: Her Majesty's Stationery Office, 1927--. Monthly. Approx. 650 pages per year.

Section 2

TECHNICAL AND TRADE JOURNALS

28 ACTA PHARMACOLOGICA ET TOXICOLOGICA. Copenhagen: Munksgaard, 1952--. Semiannual. Approx. 1,000 pages per year.

Original research papers in experimental and chemical toxicology. Reports are chiefly by Scandinavian authors or on experiments done in Scandinavian laboratories.

29 ADVANCES IN FOOD RESEARCH. Edited by Emil Mrak and George Stewart. New York: Academic Press, 1948--. Annual. Approx. 400 pages per year.

Extensive critical reviews of commodities and functional areas in food science.

30 AGRICHEMICAL AGE. Edited by R.M. Becler. Sausalito, Calif.: Jack D. Pickett, 1958--. 9 issues per year. Approx. 300 pages per year.

From 1965 to 1970 known as AGRICHEMICAL WEST. Use, cost, and effect information; new product descriptions.

31 AGRICULTURAL AND BIOLOGICAL CHEMISTRY. Tokyo: Japanese Academic Societies Center (for the Agricultural Chemical Society of Japan), 1924--. Monthly. Approx. 3,000 pages per year.

Original technical papers; important comparative journal. Text in English, French, and German.

32 AGRICULTURAL CHEMICALS. Harrisburg, Pa.: Wayne E. Dorland Publisher, 1945--. Monthly. Approx. 600 pages per year.

Semipopular style in trade journal. Products, problems, legislation, techniques described.

33 AGRICULTURAL RESEARCH. Washington, D.C.: U.S. Department of Agriculture, Agricultural Research Service, 1953--. Monthly. Approx. 180 pages per year.

Nontechnical articles dealing with current agency research on crops, pest control, livestock, nutrition, and stored products.

34 AGRICULTURAL SCIENCE REVIEW. Washington, D.C.: U.S. Department of Agriculture, Cooperative State Research Service, 1963--. Quarterly. Approx. 150 pages per year.

Critical reviews designed to provide common points of understanding among agricultural scientists. Technical reports on environmental and ecological subjects included.

35 AGRICULTURE AND ENVIRONMENT. Amsterdam: Elsevier Scientific Publishing Co., 1974--. Quarterly. Approx. 600 pages per year.

General and fascinating articles on agricultural activities worldwide. Pesticides and fertilizers frequently described.

36 AIBS BULLETIN. See BIOSCIENCE.

37 AMBIO. Oslo, Norway: Royal Swedish Academy of Sciences, 1972--. Quarterly. Approx. 180 pages per year.

Chiefly reports of environmental problems and current studies in Scandinavian countries. Semipopular style; in English.

38 AMERICAN JOURNAL OF PUBLIC HEALTH. Washington, D.C.: American Public Health Association, 1911--. Monthly. Approx. 2,500 pages per year.

Environmental, medical, and social aspects; disease prevention, vector control, and program planning.

39 AMERICAN LABORATORY. Fairfield, Conn.: International Scientific Communications, 1968--. Monthly. Approx. 1,200 pages per year.

Trade journal with advertisements, new product information, technical papers on types of equipment and analytical techniques.

40 AMERICAN SCIENTIST. New Haven, Conn.: Sigma Xi and Scientific Research Society of North America, 1913--. Bimonthly. Approx. 760 pages per year.

General scientific journal entertaining wide spectrum of topics. Currently includes ecological and environmental concerns.

41 ANALYTICAL LETTERS. New York: Marcel Dekker, 1968--. Monthly. Approx. 1,000 pages per year.

International journal making possible rapid communication among analytical chemists.

42 ANIMAL NUTRITION AND HEALTH. San Francisco: California Farmer, 1945--. Monthly. Approx. 300 pages per year.

Trade journal with new product and technical information for commercial livestock and poultry production. Includes pest control products.

43 APPLIED MICROBIOLOGY. Washington, D.C.: American Society for Microbiology, 1953--. Monthly; 2 vols. per year. Approx. 1,700 pages per year.

Applied and fundamental knowledge of microorganisms; technical level.

44 ARCHIVES OF BIOCHEMISTRY AND BIOPHYSICS. New York: Academic Press, 1942--. Monthly. Approx. 4,500 pages per year.

International journal, the standard of these disciplines. Fundamental technical level.

45 ARCHIVES OF ENVIRONMENTAL CONTAMINATION AND TOXICOLOGY. New York: Springer-Verlag, 1973--. Quarterly. Approx. 640 pages per year.

Important, full-length review articles on experimental or theoretical research.

46 ARCHIVES OF ENVIRONMENTAL HEALTH. Chicago: American Medical Association, 1960--. Monthly. Approx. 200 pages per year.

Largely medically oriented, relating to occupational health. Technical, original research reports.

47 ARCHIVES OF TOXICOLOGY. New York: Springer-Verlag, 1930--. Quarterly. Approx. 350 pages per year.

Reviews, original papers, and short communications in the full range of effects of toxic substances on animals. Includes methods of treatment, detection analysis, and forensic toxicology. Jointly titled with ARCHIV FUER TOXICOLOGIE.

48 BIOCHEMICA ET BIOPHYSICA ACTA. Amsterdam: Elsevier Publishing Co., 1947--. Irregular. Approx. 500 pages per year.

International technical journal of biochemistry and biophysics. In English.

49 BIOCHEMICAL PHARMACOLOGY. Elmsford, N.Y.: Pergamon Press, 1955--. Biweekly. Approx. 2,400 pages per year.

Standard journal reference; mainly short technical reports of original papers.

50 BIOCHEMISTRY. Washington, D.C.: American Chemical Society, 1962--. Biweekly. Approx. 4,800 pages per year.

Fundamental disciplinary standard. Technical reporting.

51 BIOSCIENCE. Arlington, Va.: American Institute of Biological Sciences, 1951--. Biweekly. Approx. 1,000 pages per year.

AIBS publication for membership. Broad range of topics and announcements. No longer publishes technical reports.

52 BULLETIN OF ENVIRONMENTAL CONTAMINATION AND TOXICOLOGY. New York: Springer-Verlag, 1966--. Monthly. Approx. 1,000 pages per year.

Research announcements in air, soil, water, and food contamination. Short reports; longer papers and reviews appear in ARCHIVES OF ENVIRONMENTAL CONTAMINATION AND TOXICOLOGY.

53 BULLETIN OF THE AGRICULTURAL CHEMICAL SOCIETY OF JAPAN. See AGRICULTURAL AND BIOLOGICAL CHEMISTRY.

54 BULLETIN OF THE ENTOMOLOGICAL SOCIETY OF AMERICA. College Park, Md.: 1955--. Quarterly. Approx. 350 pages per year.

Long articles on insect biology and control. Society information.

55 BULLETIN OF THE WORLD HEALTH ORGANIZATION. Geneva, Switzerland: World Health Organization, Office of Publications and Translations, 1947--. Bimonthly. Approx. 660 pages per year.

56 CALIFORNIA AGRICULTURE. Berkeley: University of California, Agriculture Experiment Station, 1947--. Monthly. Approx. 180 pages per year.

Progress in research by station staff and extension service. Includes articles on chemical and biological pest control.

57 CALIFORNIA'S HEALTH. Berkeley: State Department of Public Health, 1942--. Monthly. Approx. 180 pages per year.

Semipopular approach; important in part because safety of field workers in California is heavily studied. Includes studies of pesticides in specific communities.

58 CANADIAN JOURNAL OF MICROBIOLOGY. National Research Council of Canada, 1954--. Monthly. Approx. 4,800 pages per year.

Important basic journal featuring technical reports on micro-organisms.

59 CHEMICAL AND ENGINEERING NEWS. Washington, D.C.: American Chemical Society, 1923--. Weekly. Approx. 3,600 pages per year.

Semitechnical reporting and advertising in this trade journal; magazine format. World news coverage, emphasizing industrial and production phases.

60 CHEMICAL WEEK. New York: McGraw-Hill, 1914--. Weekly. Approx. 2,080 pages per year.

National and international news of the chemical industry in trade magazine format.

61 CHEMICO-BIOLOGICAL INTERACTIONS. Amsterdam: Elsevier Scientific Publishing Co., 1968--. Monthly. Approx. 1,080 pages per year.

International journal, in English, on the mechanisms by which exogenous chemicals change biological systems. More biologically than chemically oriented.

62 CHEMISTRY AND INDUSTRY. London: Society of Chemical Industry, 1963--. Biweekly. Approx. 1,200 pages per year.

Excellent informative trade journal. News reports, editorial comments, technical papers, product reviews.

63 CHEMOSPHERE. Oxford, Engl., and Elmsford, N.Y.: Pergamon Press, 1972--. Bimonthly. Approx. 360 pages per year.

Journal geared to chemical and technological solutions; industrially oriented.

64 CLINICAL TOXICOLOGY. New York: Marcel Dekker, 1968--. Quarterly. Approx. 900 pages per year.

Medically oriented. Articles involve some chemistry.

65 COMPOST SCIENCE. Emmaus, Pa.: Rodale Press, 1960--. Bimonthly. Approx. 180 pages per year.

Waste recycling; reports on the entire field of large-scale composting and recycling of organic solid wastes.

66 CONTRIBUTIONS OF THE BOYCE THOMPSON INSTITUTE FOR PLANT RESEARCH. Yonkers, N.Y.: Boyce Thompson Institute, 1925--. Quarterly. Approx. 350 pages per year.

Chiefly staff research reports in many fields of plant science--morphology, pest control, propagation, genetics, and others.

67 COOPERATIVE ECONOMIC INSECT REPORT. Hyattsville, Md.: U.S. Department of Agriculture, 1951--. Weekly. Approx. 900 pages per year.

Plant protection and quarantine programs; animal and plant health; inspection services.

68 CRITICAL REVIEWS IN ENVIRONMENTAL CONTROL. Cleveland, Ohio: CRC Press, 1970--. Quarterly. Approx. 600 pages per year.

Relatively new journal dealing with a wide range of biological, chemical, occupational, and industrial problems. Individual contributions tend to be extended reviews. CRC Press used to be the Chemical Rubber Company; many early citations refer to it that way.

69 CRITICAL REVIEWS IN TOXICOLOGY. Cleveland, Ohio: CRC Press, 1972--. Quarterly. Approx. 600 pages per year.

Extended assessments of subjects in toxicology. Often lengthy reference lists.

70 DOW DIAMOND. Midland, Mich.: Dow Chemical Co., 1938--. 3 issues per year. Approx. 100 pages per year.

An attempt to improve public relations. Some articles deal with chemicals in the environment.

71 DOWN TO EARTH. Midland, Mich.: Dow Chemical Co., 1945--. Quarterly. Approx. 120 pages per year.

Articles deal with agricultural chemicals; chiefly concerned with Dow products.

72 DRUG METABOLISM AND DISPOSITION. Baltimore, Md.: Williams and Wilkins Co., American Society for Pharmacology and Experimental Therapeutics, 1973--. Bimonthly. Approx. 540 pages per year.

Technical; biochemical reactions in living organisms.

73 DRUG METABOLISM REVIEWS. New York: Marcel Dekker, 1972--. Biannual. Approx. 240 pages per year.

Technical reports; important in toxicology.

74 ENVIRONMENT. St. Louis, Mo.: Scientist's Institute for Public Information, 1958--. Monthly. Approx. 450 pages per year.

Directs attention to environmental problems. Articles written for general public.

75 ENVIRONMENTAL CONSERVATION. Amsterdam: Elsevier Sequoia, 1974--. Quarterly. Approx. 320 pages per year.

International journal devoted to maintaining global viability through exposing and countering environmental deterioration due to unwise technology.

76 ENVIRONMENTAL EDUCATION REPORT. Washington, D.C.: Environmental Educators, 1973--. Monthly. Approx. 190 pages per year.

Coverage of funds, projects, materials, and trends in environmental education, from grass roots to global.

77 ENVIRONMENTAL HEALTH PERSPECTIVES. Research Triangle Park, N.C.: National Institute of Environmental Health, 1972--. Biannual. Approx. 1,200 pages per year.

Technical aspects of chemical problems in the environment. Major reviews and references.

78 ENVIRONMENTAL LETTERS. New York: Marcel Dekker, 1971--. Bimonthly. Approx. 400 pages per year.

International journal for rapid communication on environmental matters.

79 ENVIRONMENTAL POLLUTION. AN INTERNATIONAL JOURNAL. Essex, Engl.: Elsevier Publishing Co., 1970--. Quarterly. Approx. 1,200 pages per year.

80 ENVIRONMENTAL RESEARCH. New York and London: Academic Press, 1970--. Bimonthly. Approx. 900 pages per year.

Chemical and biological view on the effects of toxic substances on living organisms; biomedical approach.

81 ENVIRONMENTAL SCIENCE AND TECHNOLOGY. Washington, D.C.: American Chemical Society, 1967--. Monthly. Approx. 980 pages per year.

Technological, emphasizing air, water, and waste chemistry. Strongly industrially oriented.

82 FARM CHEMICALS. Willoughby, Ohio: E.G.K. Meister, 1894--. Monthly. Approx. 840 pages per year.

Magazine deals mainly with primary manufacturers, formulators, mixers, dealers, and commercial applicators of fertilizers and pesticides.

83 FDA BYLINES. See FDA CONSUMER.

84 FDA CONSUMER. Washinton, D.C.: U.S. Food and Drug Administration, 1967--. Monthly. Approx. 300 pages per year.

Official magazine of the Food and Drug Administration. Published as FDA BYLINES from 1959 to 1964, and as FDA INTERBUREAU BYLINES from 1964 to 1966.

85 FDA INTERBUREAU BYLINES. See FDA CONSUMER.

86 FEDERATION PROCEEDINGS. Bethesda, Md.: Federation of American Societies for Experimental Biology, 1942--. Monthly. Approx. 2,000 pages per year.

Contains symposia, special papers, and abstracts of papers presented at the scientific sessions of the annual meeting of the federation.

87 FOOD AND COSMETICS TOXICOLOGY. London: Pergamon Press (for the British Industrial Biological Research Association), 1963--. Bimonthly. Approx. 700 pages per year.

Technical journal related mainly to mammals; testing for human acceptances of drugs, cosmetics, and food additives.

88 FOOD CHEMICAL NEWS. Washington, D.C.: Rothschild, 1959--. Weekly. Approx. 2,080 pages per year.

Publication for executives, providing in-depth information regarding regulation of food additives, colors, pesticides, and allied products.

89 FOOD TECHNOLOGY. Chicago: Institute of Food Technologists, 1947--. Monthly. Approx. 1,200 pages per year.

Nontechnical magazine directed towards food science.

90 HEALTH LABORATORY SCIENCE. Albany, N.Y.: American Public Health Association, 1964--. Quarterly. Approx. 300 pages per year.

Discusses laboratory methods, various diseases, and improvements on testing procedures.

91 INDUSTRIAL RESEARCH. Chicago: Industrial Research, 1959--. Monthly. Approx. 900 pages per year.

Industrial research reports; management and product articles; nontechnical.

92 INTERNATIONAL JOURNAL OF ENVIRONMENTAL ANALYTICAL CHEMISTRY. London: Gordon and Breach, 1971--. Quarterly. Approx. 340 pages per year.

Original research on all aspects of analytical work related to environmental problems; analysis of organic, inorganic, and radioactive pollutants in air and water; determination of harmful substances including their metabolic breakdown products.

93 INTERNATIONAL JOURNAL OF ENVIRONMENTAL STUDIES. London: Gordon and Breach, 1970--. Quarterly. Approx. 280 pages per year.

The relationship between man and his environment.

94 INTERNATIONAL PEST CONTROL. London: McDonald Publications, 1958--. Bimonthly. Approx. 120 pages per year.

Nontechnical magazine involving pests and their controls. Includes reports on crop protection, public health, wood preservation. Important journal.

95 JAPAN PESTICIDE INFORMATION. Tokyo: Japan Plant Protection Association, 1969--. Monthly. Approx. 240 pages per year.

Magazine deals mainly with variety of pesticides used and for what purposes. In English.

96 JOURNAL OF AGRICULTURAL AND FOOD CHEMISTRY. Washington, D.C.: American Chemical Society, 1953--. Bimonthly. Approx. 1,320 pages per year.

Technical journal with heavy emphasis on analytical procedures and research results.

97 JOURNAL OF BACTERIOLOGY. Washington, D.C.: American Society for Microbiology, 1916--. Monthly. Approx. 3,900 pages per year.

Technical articles relating to all branches of fundamental microbiology.

98 JOURNAL OF CHEMICAL ECOLOGY. New York: Plenum Press, 1975--. Quarterly. Approx. 460 pages per year.

Understanding of the interactions of organisms with their environment mediated by the chemicals they produce.

99 JOURNAL OF ECONOMIC ENTOMOLOGY. College Park, Md.: Entomological Society of America, 1908--. Bimonthly. Approx. 1,800 pages per year.

Technical journal with interests based on applied aspects of insect control. Old, respected journal.

100 JOURNAL OF ENVIRONMENTAL HEALTH. Denver, Colo.: National Environmental Health Association, 1938--. Bimonthly. Approx. 650 pages per year.

Dedicated to the betterment of the nation's health through control of the environment.

101 JOURNAL OF ENVIRONMENTAL QUALITY. New York: American Society of Agronomy, 1972--. Quarterly. Approx. 560 pages per year.

Cooperation among the American Society of Agronomy, the Crop Science Society of America, and the Soil Science Society of America to bring together papers on environmental quality originating with members of these societies.

102 JOURNAL OF FOOD SCIENCE. Chicago: Institute of Food Technologists, 1936--. 7 issues per year. Approx. 1,200 pages per year.

Effects on food of various methods of handling; technical reports.

103 JOURNAL OF PHARMACOLOGY AND EXPERIMENTAL THERAPEUTICS. Baltimore, Md.: Williams and Wilkins Co., 1909--. Monthly. Approx. 2,400 pages per year.

Publication of the American Society for Pharmacology and Experimental Therapeutics. Deals with any aspect of interactions of chemicals with biological systems.

104 JOURNAL OF STORED PRODUCTS RESEARCH. Oxford, Engl.: Pergamon Press, 1965--. Quarterly. Approx. 300 pages per year.

British journal of research on advances in product storage; technical reporting.

105 JOURNAL OF THE AMERICAN CHEMICAL SOCIETY. Washington, D.C.: 1879--. Biweekly. Approx. 4,200 pages per year.

Standard source in American chemistry. Technical journal.

106 LIFE SCIENCES. New York: Pergamon Press, 1962--. Biweekly. Approx. 3,120 pages per year.

Formed by merger of LIFE SCIENCES, PART 1: PHYSIOLOGY AND PHARMACOLOGY, and LIFE SCIENCES, PART 2: BIOCHEMISTRY, GENERAL AND MOLECULAR. Technical papers covering biochemical reactions in living organisms.

107 MARINE POLLUTION BULLETIN. New York: Pergamon Press, 1970--. Monthly. Approx. 168 pages per year.

Aims to give rapid dissemination of information about the

pollution of the sea by human activities. Includes studies on the effects of effluent disposal and release of noxious materials into the sea, marine management, absorbency of the sea, natural marine ecology.

108 MICROBIAL ECOLOGY. New York: Springer-Verlag, 1974--. Quarterly. Approx. 400 pages per year.

International journal concerned with those branches of ecology in which microorganisms are involved. Articles may include significant advances in the microbiology of natural ecosystems as well as those describing new methodologies.

109 MORBIDITY AND MORTALITY. Washington, D.C.: U.S. Department of Health, Education, and Welfare, 1952--. Weekly. Approx. 500 pages per year.

Center for Disease Control, Atlanta, Georgia. Lists types and epidemiology of disease.

110 MOSQUITO NEWS. Albany, N.Y.: American Mosquito Control, 1940--. Quarterly. Approx. 700 pages per year.

Journal of the American Mosquito Control Association. Devoted to the improvement and sound extension of mosquito control and related work as a vital health service.

111 NAC NEWS AND PESTICIDE REVIEWS. Washington, D.C.: National Agricultural Chemicals Association, 1953--. Bimonthly. Approx. 100 pages per year.

Nontechnical articles reviewing the use of pesticides in agriculture. Trade journal.

112 NATIONAL WILDLIFE. Vienna, Va.: National Wildlife Federation, 1963--. Bimonthly. Approx. 400 pages per year.

The organization's statement of purpose is "to create and encourage an awareness among the people of this nation of the need for wise use and proper management of these resources of the earth upon which the lives and welfare of men depend: the soil, the air, the water, the forests, the minerals and the wildlife."

113 NATURE. London: Macmillan Journals, 1869--. Weekly. Approx. 3,900 pages per year.

Technical magazine that centers on observations or problems in the sciences. Equivalent to the American SCIENCE.

114 OFFICIAL ANALYTICAL CHEMISTS. Washington, D.C.: Association of Official Analytical Chemists, 1915--. Quarterly. Approx. 1,500 pages per year.

The journal, an old standard, is commonly known as JOAC. Most articles deal with analyses of drugs, metals, pesticides, and food contaminants.

115 PANS (PEST ARTICLES AND NEWS SUMMARIES). London: Centre for Overseas Pest Research, 1954--. Quarterly. Approx. 440 pages per year.

Purpose is to keep workers in the developing countries informed of the latest advances in all aspects of crop protection. Scope ranges in any field of pest control with particular relevance to the tropics.

116 PEST CONTROL. Cleveland, Ohio: Harvest Publishing Co., subsidiary of Harcourt Brace Jovanovich, 1933--. Monthly. Approx. 360 pages per year.

Nontechnical; primarily for practitioners in that field.

117 PEST CONTROL OPERATORS NEWS. Woodland Hills, Calif.: Bee Publishing Corp., 1941--. Monthly. Approx. 480 pages per year.

Technical publication for the professional pest control operator. Includes articles on aerial application, forestry, fumigation, golf course maintenance, mosquito abatement, ornamentals, sanitation, structural pest control, weeds and turf, vector control, vertebrate pest control, and environmental aids.

118 PESTICIDE CHEMICAL NEWS. Washington, D.C.: L. Rothschild, Jr., 1973--. Weekly. Approx. 780 pages per year.

Detailed coverage; in-depth analyses of key developments regarding laws and regulations governing pesticides and other toxic chemicals.

119 PESTICIDES. London: R.V. Raghavan, 1967--. Monthly. Approx. 720 pages per year.

Magazine describes types of pesticides, applications, and efficiency of use.

120 PESTICIDES BIOCHEMISTRY AND PHYSIOLOGY. New York: Academic Press, 1971--. Bimonthly. Approx. 570 pages per year.

Topics deal with biochemistry and physiology of insecticides, herbicides, fungicides, acaricides, and rodenticides. Also includes effects of these agents on the biochemistry and physiology of both target and nontarget organisms.

121 PESTICIDE SCIENCE. Oxford, Engl.: Blackwell Scientific Publications, 1975--. Bimonthly. Approx. 600 pages per year.

Journal of international research and technology on crop protection and pest control. Technical.

122 PESTICIDES MONITORING JOURNAL. Chamblec, Ga.: Environmental Protection Agency, Technical Services Division, 1967--. Quarterly. Approx. 250 pages per year.

Federal Working Group on Pest Management Research reports on pesticide levels and efforts. Includes Agricultural Research Service, Consumer and Marketing Service, Extension Service, Forest Service, Department of Defense, Fish and Wildlife Service, Geological Survey, FDA, EPA, National Marine Fisheries, and NSF. Initially an attempt to coordinate federal pest control decisions; now a journal open to all in pest control.

123 PHARMACOLOGICAL REVIEWS. Baltimore, Md.: Williams and Wilkins Co., 1951--. Quarterly. Approx. 500 pages per year.

Published for the American Society for Pharmacology and Experimental Therapeutics. Covers the fields of pharmacology, experimental therapeutics, and chemotherapy, discussing how various drugs can be harmful and how they work.

124 PHYTOPATHOLOGY. Worcester, Mass.: Hefferman Press, 1911--. Monthly. Approx. 2,000 pages per year.

International journal of the American Phytopathological Society. Technical reports on plant pathology.

125 POULTRY SCIENCE. Guelph, Ontario, Canada: Poultry Science Association, 1908--. Bimonthly. Approx. 2,500 pages per year.

Official journal of the Poultry Science Association. Technical; useful not only for poultry but as background for all bird studies.

126 PROCEEDINGS OF THE SOCIETY FOR ANALYTICAL CHEMISTRY. London: Chemical Society, 1964--. Monthly. Approx. 350 pages per year.

Meeting reports and technical research articles dealing with chemistry.

127 PYRETHRUM POST. Nehuru, Kenya: Pyrethrum Bureau, 1948--. Semi-annual. 180 pages per year.

Announcements on uses, sales, and effects of pyrethrum.

128 RESIDUE REVIEWS. RESIDUES OF PESTICIDES AND OTHER FOREIGN CHEMICALS IN FOODS AND FEEDS. New York: Springer-Verlag, 1962--. Irregular. Approx. 150 pages per vol.; 1-3 vols. per year.

Excellent review articles in each volume; chiefly pesticide matters. Increasingly wider pollutant coverage. Frequency is dependent on contributions. Treated as a journal for citation.

129 REVIEWS ON ENVIRONMENTAL HEALTH. London: Freund Publishing House, 1974--. Quarterly. Approx. 400 pages per year.

Intended to fill the need for short, highly specialized reviews in the field of environmental health and pollution control.

130 RUSSIAN PHARMACOLOGY AND TOXICOLOGY. London: Euromed Publications, 1938--. Bimonthly. Approx. 350 pages per year.

English version of the leading Russian pharmacological journal FARMAKOLOGIYA I. TOKSIKOLOGIYA--an organ of the Soviet Ministry of Health.

131 SCIENCE. Washington, D.C.: American Association for the Advancement of Science, 1880--. Weekly. Approx. 3,900 pages per year.

Technical magazine covering various topics within science, ranging anywhere from biological excretions to early diggings.

132 THE SCIENCE OF THE TOTAL ENVIRONMENT. Amsterdam: Elsevier Scientific Publishing Co., 1972--. Quarterly. Approx. 450 pages per year.

International technical journal for scientific research into the environment and its relationship with man. In English.

133 SCIENTIFIC AMERICAN. New York: Scientific American, 1845--. Monthly. Approx. 2,400 pages per year.

Basic science on wide range of subjects; written for general, educated audiences.

134 SPAN. London: Shell International Chemical Co., 1958--. 3 issues per year. Approx. 200 pages per year.

Title is acronym for Shell Progress in Agriculture News. Written for layperson. Covers many aspects of agriculture, from pest control to policies.

135 TOXICOLOGICAL AND ENVIRONMENTAL CHEMISTRY REVIEWS. London: Gordon and Breach Science Publishers, 1972--. Quarterly. Approx. 320 pages per year.

International review journal devoted to the fundamental aspects of analysis, metabolism, and general chemistry and biochemistry of xenobiotic compounds and natural toxins as related to the environment and human health.

136 TOXICOLOGY. Amsterdam and New York: Elsevier/North-Holland, 1973--. Bimonthly. Approx. 780 pages per year.

Describes biological effects of administration of chemical compounds, principally to animals, but also to man. Standard technical reference.

137 TOXICOLOGY AND APPLIED PHARMACOLOGY. New York and London: Academic Press, 1959--. Monthly. Approx. 1,200 pages per year.

Official journal of the Society of Toxicology. Scientific research pertaining to alterations in tissue structure or function resulting from the administration of chemicals, drugs, or natural products to animal or man.

138 VETERINARY TOXICOLOGY. Manhattan, Kans.: American College of Veterinary Toxicologists, 1958--. Quarterly. Approx. 200 pages per year.

Nontechnical, society publication.

139 WAR ON HUNGER. Washington, D.C.: Agency for International Development, Publications Division, Office of Public Affairs, 1967--. Monthly. Approx. 180 pages per year.

General contributions for and from those in AID whose purpose is to increase food production and services in the developing countries.

140 WATER POLLUTION CONTROL. Washington, D.C.: Water Pollution Control Federation, 1928--. Monthly. Approx. 3,000 pages per year.

Technical articles involving water pollution. Professional journal known as SEWAGE WORKS JOURNAL until 1950.

141 WEED RESEARCH. Oxford, Engl.: Blackwell Scientific Publications, 1961--. Bimonthly. Approx. 400 pages per year.

Official journal of the Environmental Weed Research Council. European equivalent of the American journal WEED SCIENCE. Plant pest control.

142 WEED SCIENCE. Champaign, Ill.: Weed Science Society of America, 1951--. Bimonthly. Approx. 600 pages per year.

Technical magazine; plant pest control.

143 WORLD REVIEW OF PEST CONTROL. London: World Review of Pest Control, 1962--. Quarterly. Approx. 200 pages per year.

Global coverage of news and technical information. Written at professional level.

144 YOUR ENVIRONMENT. London: Your Environment, 1969--. Quarterly. Approx. 200 pages per year.

Current environmental causes, concerns, and corrections. Written at popular level.

145 XENOBIOTICA. THE FATE OF FOREIGN COMPOUNDS IN BIOLOGICAL SYSTEMS. London: Taylor and Francis, 1971--. Annual. Approx. 600 pages per year.

Topics include pharmaceuticals, drugs, pesticides, additives. International contributors.

Section 3

BACKGROUND REFERENCES IN FIVE RELATED FIELDS

Sections 1 and 2 dealt with a broad range of journal and other sources and the means of retrieval of information from them. Section 3 narrows the field of references to five areas of environmental toxicology of particular interest to the environmentalist. All references in this section are secondary sources or background papers.

ECOLOGY

Every effect of a pollutant in the environment is an ecological interaction. A pollutant is a chemical substance which produces undesirable effects on living things. These effects may be studied in many ways. Toxicologists normally follow effects within the organism; chemists may do the same but frequently follow chemical pathways both inside and outside the organism. Ecologists tend to view foreign chemicals in ecosystems in two ways. The first is to study the effects of such a chemical or chemicals on favored groups of organisms such as a crop species, birds, fish, game, or a particular plant community. The professional ecologist is more apt to describe effects of chemicals as they interact at given trophic levels and as these effects are telegraphed along subsequent levels in ecosystems. The environmental toxicologist must incorporate in his research and manner of organization all the foregoing approaches. Whatever his discipline and speciality, he must think as an ecologist does. The titles listed below constitute only a sampling of the range of problems and emphases in the ecological arena.

146 Anderson, Daniel W., and Hickey, Joseph J. "Eggshell Changes in Certain North American Birds." PROCEEDINGS OF THE XVth INTERNATIONAL ORNITHOLOGICAL CONGRESS 1972, pp. 514-40.

Data-filled review of the thinning process in bird eggs based on over 35,000 measurements. Most are of pre-1947 eggs, before the onset of general environmental contamination with chlorinated hydrocarbons. Eleven species are compared. Eggshell changes are rare before 1939 and common later. These species have sustained shell changes (reductions in weight or thickness) of 20 percent or more, at least for brief periods: peregrine falcon in at least three regions; marsh hawk and brown pelicans in two regions; and the prairie falcon, Cooper's hawk, double-crested cormorant, black-crowned night heron, bald eagle, and osprey in at least one region each. Regional declines in eggshell thickness in eight cases. Some declines continue. Excellent source.

147 ______. "Oological Data on Egg and Breeding Characteristics of Brown Pelicans." WILSON BULLETIN 82 (1970): 14-28.

Important background information to understand the remarkable degree of eggshell thinning and local population declines in brown pelicans, due to chlorinated hydrocarbon residues in the marine environment.

148 Baker, Herbert G. "The Evolution of Weeds." ANNUAL REVIEW OF ECOLOGY AND SYSTEMATICS 5 (1974): 1-24.

Good review; includes 191 references. In his introduction Baker states that a plant is a weed "if, in any specified geographical area, its populations grow entirely or predominantly in situations markedly disturbed by man."

149 Bray, J.R., and Gorham, E. "Litter Production in the Forests of the World." ADVANCES IN ECOLOGICAL RESEARCH 2 (1964): 101-57.

Organic forest debris considered. Components, net production, and factors affecting debris are discussed.

150 Colwell, R.R., and Morita, R.Y., eds. EFFECT OF THE OCEAN ENVIRONMENT ON MICROBIAL ACTIVITIES. Baltimore, Md.: University Park Press, 1974. 604 p. Illus.

Based on U.S.-Japan Conference on Marine Microbiology. Many contributors to this background book; not directly related to marine pollution.

151 Craig, Robert B., and Rudd, Robert L. "The Ecosystem Approach to Toxic Chemicals in the Biosphere." In SURVIVAL IN TOXIC ENVIRONMENTS, edited by M.A.Q. Khan and John P. Bederka, Jr., pp. 1-24. New York: Academic Press, 1974.

Background data on the Clear Lake and Central Valley (California) disturbed ecosystems. Presents computer-assisted modeling of aquatic and terrestial ecosystems into which pesticide chemicals intrude.

152 Daubenmire, R. "Ecology of Fire in Grasslands." ADVANCES IN ECOLOGICAL RESEARCH 5 (1968): 209-66.

Reports on fire as a major factor in composition, nutrient cycling, nutritive base for herbivores, and regional productivity. Good literature list.

153 DeVos, A. "Ecological Conditions Affecting the Production of Wild Herbivorous Mammals on Grasslands." ADVANCES IN ECOLOGICAL RESEARCH 6 (1969): 137-83.

Protein scarcity is emphasis of article. Compares productivity, chiefly African and North American, some Russian. Good review of practical side.

154 Elton, Charles S. THE ECOLOGY OF INVASIONS BY ANIMALS AND PLANTS. London: Methuen and Co., 1958. 181 p.

Readable account by world's best-known ecologist. Basic analysis in semipopular style on movement and establishment of pest species. Based on a British Broadcasting Corporation lecture series.

155 _____. THE PATTERN OF ANIMAL COMMUNITIES. New York: John Wiley & Sons, 1966. 432 p.

Two major contributions: (1) summation of twenty years of observations by many ecologists at a single place; (2) extrapolation to all ecosystems because of organized manner of classification. Elton concludes that systems are difficult to describe, and energy relationships yet harder.

156 Engelmann, Manfred D. "Energetics, Terrestrial Field Studies and Animal Productivity." ADVANCES IN ECOLOGICAL RESEARCH 3 (1966): 73-115.

Status report and review of the trophodynamic concept; important because most studies of energetics focus on aquatic ecosystems. This study summarizes terrestrial energetics. Excellent.

157 Farnworth, Edward G., and Golley, Frank B., eds. FRAGILE ECOSYSTEMS. EVALUATION OF RESEARCH AND APPLICATIONS IN THE NEOTROPICS. New York: Springer-Verlag, 1973. xxvi, 258 p. Paper.

A report of the Institute of Ecology, University of Georgia, Athens. Thoughtful appraisal of tropical productivity and potential. Reports are prepared by working parties of experienced ecologists. Strong basic yet practical emphasis. Many references.

158 Florkin, Marcel, and Scheer, Bradley T., eds. CHEMICAL ZOOLOGY.

9 vols. New York: Academic Press, 1974. xxiii, 537 p. Over 6,000 p. in series.

This valuable reference is composed of phylogenetically arranged contributions by individual specialists. For example, volume 1, PROTOZOA, has fourteen contributors; volume 9, AMPHIBIA AND REPTILIA, has twenty.

159 Gates, David M. "Toward Understanding Ecosystems." ADVANCES IN ECOLOGICAL RESEARCH 5 (1968): 1-35.

Toward understanding plants, energy, and environment. Good questions asked and reviewed regarding complexity of ecosystems. Good references.

160 Giles, Robert H., ed. WILDLIFE MANAGEMENT TECHNIQUES. 3d ed. Washington, D.C.: Wildlife Society, 1969. 623 p.

Twenty-four chapters; different contributors. Covers all phases of observation, marking, analyzing, and manipulation of populations of chiefly game birds and mammals.

161 Golterman, H.L., and Clymo, R.S., eds. CHEMICAL ENVIRONMENT IN THE AQUATIC HABITAT. Amsterdam: N.V. Noord Hollandsche Uitgevers Maatschappii, 1967. 322 p. Paper.

Proceedings of International Biological Program meeting in 1966. Many papers, chiefly limnological, but some related to environmental toxicology.

162 Goodman, Gordon T., et al., eds. ECOLOGY AND THE INDUSTRIAL SOCIETY. New York: John Wiley & Sons, 1965. viii, 395 p. Graphs, tables.

A symposium of the British Ecology Society, 13-16 April 1964. Sixteen contributors on all major aspects of environmental pollution. American, German, and English authors.

163 Gulland, J.A. "Ecological Aspects of Fishery Research." ADVANCES IN ECOLOGICAL RESEARCH 7 (1971): 115-76.

Practically directed review on dynamics of populations and various types of predictive modeling. Long reference list.

164 Hickey, Joseph J., ed. PEREGRINE FALCON POPULATIONS: THEIR BIOLOGY AND DECLINE. Madison: University of Wisconsin Press, 1969. xxii, 596 p.

Excellent summaries and papers resulting from a conference held in Madison in late 1965. Most information concerns numbers and distribution, but pesticide residues are considered. This is the best reference on the peregrine falcon, a highly endangered species.

165 Hill, Elwood F., et al. LETHAL DIETARY TOXICITIES OF ENVIRONMENTAL POLLUTANTS TO BIRDS. Special Scientific Report--Wildlife, no. 191. Washington, D.C.: Fish and Wildlife Service, Bureau of Sports Fisheries and Wildlife, 1975. 61 p.

Compilation and analysis of the results of testing 131 chemicals in bobwhite quail, Japanese quail, ring-necked pheasants, and mallard ducks. Extensive tables and appendixes.

166 Holling, C.S. "Resilience and Stability of Ecological Systems." ANNUAL REVIEW OF ECOLOGY AND SYSTEMATICS 4 (1973): 1-23.

Holling suggests that acceptance of resilience idea makes possible qualitative predictions in managed systems.

167 Hopkirk, John D. "Endemism in Fishes of the Clear Lake Region of Central California." UNIVERSITY OF CALIFORNIA PUBLICATIONS IN ZOOLOGY 96 (1973): 1-135. 4 plates.

Background for those wishing close detail on the well-studied Clear Lake pesticide concentration situation.

168 Krebs, Charles J., and Myers, Judith H. "Population Cycles in Small Mammals." ADVANCES IN ECOLOGICAL RESEARCH 8 (1974): 267-399.

Population regulation must be under evolutionary control. Extended paper concerns voles and lemmings particularly but is applicable in part to all cycling populations.

169 Kreitzer, J.F. "Eggshell Thickness in Mourning Dove Populations." JOURNAL OF WILDLIFE MANAGEMENT 35 (1971): 563-64.

Thickness of eggshells of doves (*Zenaidura macroura*) did not differ in samples from 1969-70 compared to those taken in the period 1861-1935. Thickness therefore does not respond to environmental contaminants as it does with many carnivorous birds.

170 Langford, Arthur N., and Buell, Murray F. "Integration, Identity and Stability in the Plant Association." ADVANCES IN ECOLOGICAL RESEARCH 6 (1969): 83-135.

Recent view of an old argument in plant ecology--the nature of the community.

171 LeCren, E.D., and Holdgate, M.W., eds. THE EXPLOITATION OF NATURAL ANIMAL POPULATIONS. New York: John Wiley & Sons, 1962. xiv, 399 p.

Symposium number two, British Ecology Society, 28-31 March 1960. Reports of population ecologists on both practical and theoretical aspects of exploitation of fishes, birds, large game mammals, whales, and seals. International representation. Still an excellent source.

172 MacFayden, A. "Soil Arthropod Sampling." ADVANCES IN ECOLOGICAL RESEARCH 1 (1962): 1-34.

Practical summary of current methods. Concerns three types of studies in ecology: exploratory, community, and trophic.

173 Mann, K.H. "The Dynamics of Aquatic Ecosystems." ADVANCES IN ECOLOGICAL RESEARCH 6 (1969): 1-81.

Major review of the trophic concept. Many graphs and models. Essentially a quarter-century review since Raymond L. Lindemann's classic 1942 work (ECOLOGY 23: 399-418). Long reference list.

174 Mathews, J.G., comp. GLOSSARY OF AQUATIC ECOLOGICAL TERMS. Rev. ed. Ada, Okla.: Environmental Protection Agency, Region XI, 1972. 82 p.

Proper technical description of terms commonly used in aquatic ecology. Directed particularly to those concerned with water pollution.

175 Miller, Richard S. "Pattern and Process in Competition." ADVANCES IN ECOLOGICAL RESEARCH 4 (1967): 1-74.

Unusually good review of process of competition, niche concept, competitive exclusion, and species diversity. Includes vertebrates and invertebrates. Long reference list.

176 Murdoch, W[illiam]. W., and Oaten, A. "Predation and Population Stability." ADVANCES IN ECOLOGICAL RESEARCH 9 (1975): 1-131.

Major technical review on predation, a subject that should be referred to by anyone interested in population regulation by any means. Much original material; many references.

177 Odum, Eugene P. FUNDAMENTALS OF ECOLOGY. 3d ed. Philadelphia: W.B. Saunders Co., 1971. xiv, 514 p. Graphs, photographs, models.

Basic; best-known text. Many practical applications and implications.

178 _____. "The Strategy of Ecosystem Development." SCIENCE 164 (1969): 262-70.

Provides an understanding of ecological structuring. Ecosystemic studies are essential to resolving man's conflicts with nature. "Must" reading.

179 Ovington, J.D. "Quantitative Ecology and the Woodland Ecosystem Concept." ADVANCES IN ECOLOGICAL RESEARCH 1 (1962): 103-92.

Stresses fundamental unity of ecosystem physiology. Practical implications in pollution. Long reference list.

180 Pimentel, David, ed. INSECTS, SCIENCE AND SOCIETY. New York: Academic Press, 1975. xxv, 284 p.

Papers from a conference at Cornell University in October 1974, honoring pioneer entomologist John Henry Comstock. Contributions unusually good. Sections are (1) "Insects and Their Social Implications"; (2) "Communication among Insects"; (3) "Patterns and Processes in Insect Interactions"; (4) "Insect Population Dynamics"; and (5) "Insect-Pest Management."

181 Raymont, J.E.G. "The Production of Marine Plankton." ADVANCES IN ECOLOGICAL RESEARCH 3 (1966): 117-205.

Conversion in sea to high potential chemical energy comes from autotrophs--bacteria or green plants. The latter are overwhelmingly more important. Both phytoplankton and zooplankton productivity analyzed. Basic to understanding marine pollution.

182 Smith, D.F., and Hill, D.M. "Natural and Agricultural Ecosystems." JOURNAL OF ENVIRONMENTAL QUALITY 4 (1975): 143-45.

Excellent, thoughtful review and argument. The well-conceived thesis is that awareness of environmental problems cannot properly be contained by merely bringing different research papers together within the framework of older scientific disciplines. A new consciousness is required. For example, no distinction should be made between natural and agricultural ecosystems; they are simply different expressions along a biophysical continum.

183 Stickel, William H. "Some Effects of Pollutants in Terrestrial Ecosystems." In ECOLOGICAL TOXICOLOGY RESEARCH, edited by A.D. McIntyre and C.F. Mills, pp. 25-74. New York: Plenum Publishing Corp., 1975.

Excellent review centering on organochlorines and heavy metals. Long reference list.

184 Streifer, William. "Realistic Models in Population Ecology." ADVANCES IN ECOLOGICAL RESEARCH 8 (1974): 200-266.

Streifer presents models based on physiological, ecological, and social behavior of individuals in populations. Models provide insight and have predictive validity. They are useful in determining optimal strategies for pest control, harvesting, preservation of species, and so forth.

185 Van Dyne, George M., ed. THE ECOSYSTEM CONCEPT IN NATURAL RESOURCE MANAGEMENT. New York: Academic Press, 1969. 383 p.

Based on a symposium at the annual meeting of the American Society of Range Management, Albuquerque, New Mexico, 12-15 February 1968. Includes contributions from various management fields--range, forest, watershed, fishery, wildlife. Important because it provides a comparative evaluation of the ecosystem concept in several resource management fields.

186 Watt, Kenneth E.F. ECOLOGY AND RESOURCE MANAGEMENT. New York: McGraw-Hill, 1968. xii, 450 p.

Basic review text by an eminent ecologist. On the practical side but with broad perspective. Simple mathematical models.

187 White, Clayton M. "Diagnosis and Relationships of the North American Tundra-inhabiting Peregrine Falcons." AUK 85 (1968): 179-97.

Background paper on variation and systematics. Falcon response to poisoning by environmental contaminants is of special significance here.

188 White, Lynn, Jr. "The Historical Roots of Our Ecologic Crisis." SCIENCE 155 (1967): 1203-7.

A thoughtful review of the origins of scientific and technological applications as they relate to environmental disruptions today.

189 Wood, J.M. "Biological Cycles for Toxic Elements in the Environment." SCIENCE 183 (1974): 1049-52.

Toxic substances are either naturally occurring or industrially synthesized. Naturally occurring toxins may be hazardous if their distribution, normally relatively constant, becomes unbalanced. Only when substances enter industrial processes and reenter the environment are balances upset. An example is methyl mercury. Table 1 lists elements according to their toxicity, and their envrionmental accessibility. Predictions about pollutants can be made based on their reactive capacities. Models for mercury and arsenic accompany the article.

190 Woodwell, George M. "Toxic Substances and Ecological Cycles." SCIENTIFIC AMERICAN 216 (1967): 24-31.

An excellent general introduction to the ecological manner by which persistent chemicals (radionuclides and pesticides) distribute themselves in the environment and how they may concentrate to dangerous levels.

ENVIRONMENTAL QUALITY AND SAFETY

Both intentional and accidental release of chemicals into the environment must be judged by the degree of risk to human values which their presence entails. The words "quality" and "safety" are difficult to define precisely. Qualitative judgments often depend on the values and premises of the one doing the judging. Disagreement is inevitable and much controversy over environmental quality stems from this subjectivity. Safety standards are more definable but here, too, there are arguable differences in values and unknowns that make understanding and regulation difficult.

191 Borlaug, Norman I. "Mankind and Civilization at Another Crossroad. In Balance with Nature---A Biological Myth." BIOSCIENCE 22 (1972): 41-43.

Borlaug--"father of the green revolution"--won the Nobel Peace Prize in 1970. This article shows that even so distinguished a person can be as emotional, dogmatic, and "selective" as he accuses others (notably Rachel Carson) of being.

192 Cave, A.J., ed. PROCEEDINGS OF THE FIRST INTERNATIONAL CONGRESS OF ECOLOGY: STRUCTURE, FUNCTIONING AND MANAGEMENT OF ECOSYSTEMS. Wageningen, Holland: Centre for Agricultural Publishing and Documentation, 1974. 414 p.

Congress at The Hague in September 1974. Remarkably varied topics by well-known ecologists.

193 Council on Environmental Quality. ENVIRONMENTAL QUALITY. Washington, D.C.: Government Printing Office, 1969--. Annual.

The council, an advisory group to the president of the United States, reviews major questions of land use, resources, and pollution. The latest report, over 600 pages long, appeared in 1974.

194 Farvar, M. Taghi, and Milton, John P., eds. THE CARELESS TECHNOLOGY: ECOLOGY AND INTERNATIONAL DEVELOPMENT. New York: Natural History Press, 1972. xxix, 1,030 p.

Record of a 1968 conference in Warrenton, Virginia, on the theme of ecological aspects of international development. Begins with summary by Barry Commoner of "ecological failures." Contains six sections with fifty papers covering wide range of topics. Approximately twelve articles concern environmental toxicology directly.

195 Gough, Beverley E., ed. WORLD ENVIRONMENTAL DIRECTORY. Silver Spring, Md.: Business Publishers, 1975. cxxvi, 811 p.

Detailed listing of companies, products, and publications.

196 Massachusetts Institute of Technology [Sponsor]. MAN'S IMPACT ON THE GLOBAL ENVIRONMENT: ASSESSMENT AND RECOMMENDATIONS FOR ACTION. Cambridge, Mass.: MIT Press, 1970. xxii, 319 p.

Under the SCEP (Study of Critical Environmental Problems) program over 100 professionals contributed to this workshop report (in 1969-70) resulting from several weeks of presentations and interchange. Covers many broad aspects but includes DDT, the heavy metals, oils, wastes, monitoring concepts, production and protection aspects, and estuarine pollution.

197 Murdoch, William W., ed. ENVIRONMENT--RESOURCES, POLLUTION AND SOCIETY. 2d ed. Sunderland, Mass.: Sinauer Associates, Publishers, 1975. viii, 488 p.

Intended as a text, but is more than that. Nineteen chapters written by authorities in their individual fields. Written and illustrated for a general audience, with good referencing. See particularly Murdoch's opening and closing chapters. Strong social themes throughout.

198 Pimentel, David, et al. "Food Production and the Energy Crisis." SCIENCE 182 (1973): 443-49.

Impressive review of the relationship among crop production, world food demands, and energy inputs including petroleum products, fertilizers, insecticides, and herbicides. "High-energy" crops are exemplified by corn production. Both insecticidal and herbicidal use continues to increase.

199 President's Science Advisory Committee. RESTORING THE QUALITY OF OUR ENVIRONMENT. Washington, D.C.: The White House, 1965. xii, 317 p. Paper.

Report of the Environmental Pollution Panel of the President's Science Advisory Committee. This is one of the earlier surveys of pollution sources and effects, but brings in the idea that values must be considered. Some excellent recommendations still valid.

200 Rudd, Robert L. "Chemicals in the Environment." CALIFORNIA MEDICINE 113 (1970): 27-32.

A general review of the intrusion of synthetic chemicals in the environment, with emphasis on organochlorine, organomercurial, and lead compounds.

201 Sax, N. Irving, ed. DANGEROUS PROPERTIES OF INDUSTRIAL MA-

TERIALS. 4th ed. New York: Van Nostrand Reinhold Co., 1975. 1,258 p.

Twelve sections by different contributors. Longest and most useful is section 12, "General Chemicals," by Sax. In this section many pollutants are listed by chemical name. Properties and toxicities of pollutants are given.

202 Schubert, Jack. "A Program to Abolish Harmful Chemicals." AMBIO 1 (1972): 79-89.

Schubert states that tests to detect harmful effects of chemicals are grossly inadequate. The article makes several points and suggestions: (1) it is unrealistic to expect voluntary compliance; compliance must be legislated; (2) penalties for violation should be realistically punitive; (3) noncompliancy and violation should be publicized widely; (4) long-range genetic and toxicity testing should be done in special laboratories with cooperation and shared expense by industry and government, with public representation.

203 Stoddurd, Gloria J., comp. COMMON ENVIRONMENTAL TERMS. Washington, D.C.: Environmental Protection Agency, 1973. 22 p.

Nontechnical definitions of several hundred terms associated with environmental problems. Useful. Definition of toxicity, for example: "The quality or degree of being poisonous or harmful to plant or animal life."

204 Water Resources Scientific Information Center. ESTUARINE POLLUTION: A BIBLIOGRAPHY. Washington, D.C.: Government Printing Office (for Office of Water Resources Research), U.S. Department of Interior, 1972. 477 p.

Precise referencing and indexing. In four sections: (1) significant descriptive index; (2) bibliography; (3) comprehensive index; and (4) author index.

205 Wilson, William K., et al., eds. WORLD DIRECTORY OF ENVIRONMENTAL RESEARCH CENTERS. 2d ed. New York: Oryx Press, 1974. xi, 330 p.

Sponsored by Lake Erie Environmental Program (LEES). A useful directory of agencies and institutions.

206 Worden, A.N. "Man, Animal and the Environment, More Especially in Wales." INTERNATIONAL JOURNAL OF ENVIRONMENTAL STUDIES 7 (1975): 107-17.

A general but sophisticated essay on how we equate human needs and sensitivities with productive realities.

207 World Health Organization. HEALTH HAZARDS IN THE HUMAN ENVIRONMENT. Rev. ed. Geneva: 1973. 387 p. Paper.

Relates to human health. Prepared by 100 specialists in fifteen countries. Part 2 concerns environmental pollutants including heavy metals, chlorinated hydrocarbons (DDT, PCB), nitrates, and nitrites. Much data included on laboratory and domestic animals. Long reference lists. Chapter 13 is on the special problems of mutagens, carcinogens, and teratogens.

PEST CONTROL

Chemicals may be deliberately introduced into the environment to lower the population numbers of organisms that are nuisances, that cause disease in man or domestic animals, or that interfere with economical food and fiber production. Historically, toxic chemicals have been used for these purposes. These continue in use but in recent years other methods have come into play whenever possible. These include biological control, physical manipulation, and sex attractrants. The control of noxious organisms is basically an exercise in manipulating population dynamics. Control requires a great deal of knowledge of basic population biology before effective manipulation can be practiced. Pest control has immense importance to man. One must realize that chemicals are applied only to shift biological relationships. The following is a sampling of references that cover this important field.

208 Geier, P.W., and Springett, B.P. "The Context of Pest Control." AGRICULTURE AND ENVIRONMENT 1 (1974): 373-83.

Critical review of three broad approaches to pest control: opportunistic, gradualistic, and holistic. All approaches are deficient. Authors suggest that pest control is a special aspect of bioengineering which could be termed "protection ecology." Control should be separated administratively and in teaching, research, and advisory aspects.

209 Graham, Frank, Jr. SINCE SILENT SPRING. Boston: Houghton Mifflin Co., 1970. xiv, 333 p.

This is both a sequel to Rachel Carson's SILENT SPRING and an analysis of the pesticide controversy in the 1960s. Not a technical book but has great social meaning.

210 Gunn, D.L. "Dilemmas in Conservation for Applied Biologists." ANNALS OF APPLIED BIOLOGY (London) 72 (1972): 105-27.

Address of the president of the Association of Economic Biologists, 6 July 1972. Concerned mainly with DDT in practical use and the arguments and facts employed by conservationists against its use. A thoughtful paper on a subject often polemic.

211 Headley, Joseph Charles. "Economics of Agricultural Pest Control." ANNUAL REVIEW OF ENTOMOLOGY 17 (1972): 273-86.

View by agricultural economist on several ways of evaluating cost-benefit ratios. Eighteen references.

212 McMillen, Wheeler. BUGS OR PEOPLE? New York: Appleton-Century-Crofts, 1965. xi, 228 p.

A brief, well-written but general statement about the value of pesticides. Not stated, but clearly a rebuttal to Rachel Carson's SILENT SPRING (see section 4).

213 Metcalf, Robert L., ed. ADVANCES IN PEST CONTROL. New York: Interscience Publishers, 1957--. Annual.

Annual volumes with contributed papers. Excellent quality of technical papers.

214 Rabb, R.L., and Guthrie, F.E., eds. CONCEPTS OF PEST MANAGEMENT. Raleigh: North Carolina State University, 1970. xi, 242 p.

Proceedings of a conference held at North Carolina State University, Raleigh, 25-27 March 1970. Fifteen excellently qualified contributors on special topics. Note particularly the following: T.R.G. Southwood and M.J. Way, "Ecological Background to Pest Management"; K.E.F. Watt, "The Systems Point of View in Pest Management"; R.F. Smith, "Pesticides--Their Use and Limitations in Pest Management"; David Pimental, "Training in Pest Management and the Systems Approach to Control."

The twenty-three nations and 300 participants at the conference attest to the global significance of the problem.

215 Reitz, Louis P., ed. BIOLOGICAL AND CHEMICAL CONTROL OF PLANT AND ANIMAL PESTS. Publication no. 61. Washington, D.C.: American Association for the Advancement of Science, 1960. xii, 273 p.

A symposium presented by Section O on Agriculture at the Indianapolis meeting of the AAAS, 28-30 December 1957. Nineteen contributors; somewhat superseded by later data but still useful. In three sections: (1) "The Public's Stake in Pest Control"; (2) "Recent Advances in Chemical Control"; (3) "Biological Control of Pests."

216 Rockstein, Morris. THE PHYSIOLOGY OF INSECTA. 6 vols. New York: Academic Press, 1973-74.

Forty-five contributions from individual investigators throughout the six volumes describe a wide range of topics in insect biology. Basic technical information essential for full understanding of insect toxicology.

217 Stern, Vernon M. "Economic Thresholds." ANNUAL REVIEW OF ENTOMOLOGY 18 (1973): 259-80.

Review of concepts involved in relating pest densities to costs of control measures. Incorporates all measures of "cost," including social and biological. Suggests that present methods are empirical and that more sophisticated methods are needed by applied population ecologists. Includes 104 references.

218 Van Emden, Helmut F. PEST CONTROL AND ITS ECOLOGY. Institute of Biology's Studies in Biology, no. 38. London: Edward Arnold, 1974. 60 p.

Brief, general account of several major insect pest control methods. Broad ideas illustrated.

219 Wood, R.K.S., ed. BIOLOGICAL PROBLEMS ARISING FROM THE CONTROL OF PESTS AND DISEASES. Symposium of the Institute of Biology, no. 9. London: Church Army Press, 1960. viii, 116 p.

Thirteen papers plus discussions by well-known authors. Valuable because of its thoughtful approach, easy style of presentation, and its historical position at the midpoint of major doubts about pesticide and drug effects.

The entire Institute of Biology series of publications is useful to students and interested general public; all publications are authoritative.

220 Wright, Jonathan W., et al. "Changing Concepts of Vector Control in Malaria Eradication." ANNUAL REVIEW OF ENTOMOLOGY 17 (1972): 75-102.

A review of global control methods. Various approaches to vector control employ chemical, biological, genetic, or physical means. Alternative chemicals are needed; nonchemical means cannot yet be generally applied. Epidemiological summation. Seventy-three references. World Health Organization report.

POLLUTION

One definition of a weed is a plant in the wrong place. By analogy, pollution can be defined as a chemical in the wrong place. Moreover, it must be present in an amount sufficient to produce an unwanted effect. Any material can be a pollutant. Even water at the wrong temperature can be so considered. For our purposes, however, most pollutants should be considered as single chemicals normally derived from man's activities.

221 Barrakette, Eural S., ed. POLLUTION: ENGINEERING AND SCIENTIFIC SOLUTIONS. New York: Plenum Press, 1973. xv, 784 p.

Published proceedings of the first international meeting of the Society of Engineering Science, Tel Aviv, Israel, 12-17 June 1972. Extensive contributions on all aspects, technical and social, of environmental pollution.

222 Bond, Richard G., and Straub, Conrad P., eds. CRC HANDBOOK OF ENVIRONMENTAL CONTROL. 5 vols. Cleveland, Ohio: CRC Press, 1973-74.

Urban-centered and not necessarily toxicological, but each volume has something on biological or ecological effects on animals and plants. Text consists chiefly of tables; long reference lists. Good for technical data summaries.

Individual titles are: Vol. 1, AIR POLLUTION; Vol. 2, SOLID WASTES; Vol. 3, WATER SUPPLY AND TREATMENT; Vol. 4, WASTE WATER: TREATMENT AND DISPOSAL; Vol. 5, HOSPITAL AND HEALTH CARE FACILITIES. 3,321 p.

223 Cooke, A.S. "Shell Thinning in Avian Eggs by Environmental Pollutants." ENVIRONMENTAL POLLUTION 4 (1973): 85-152.

An important major review of the phenomenon of eggshell thinning. Declines in shell thickness since World War II are attributable to residues of p,p'-DDE or other metabolites of the DDT group. In North America, PCBs and cyclodienes (notably dieldrin) play minor roles but are more important in Great Britain. Mercury compounds apparently do not figure in shell thinning. Supported by laboratory evidence, indicating interspecific differences. Gallinaceous birds least susceptible, falcons most. No single biochemical mechanism explains thinning.

224 Crosby, Donald G. "The Toxicant Wildlife Complex." PURE AND APPLIED CHEMISTRY 42 (1975): 233-51.

Direct exposure to toxicants is more prevalent than anticipated. Bioconcentration of persistent compounds may be approximated by their lipid-water partition/coefficients. Crosby suggests that biomagnification does not inevitably occur but that prediction of such effects is becoming increasingly precise.

225 Cummins, Kenneth W. "Trophic Relations in Aquatic Insects." ANNUAL REVIEWS OF ENTOMOLOGY 18 (1973): 183-206.

Excellent review of biological community construction in freshwater (primarily North American) ecosystems. Tables, graphs, charts. Seventy-two references. Background for water pollution studies.

226 Dugan, Patrick R. BIOCHEMICAL ECOLOGY OF WATER POLLUTION.

New York: Plenum Press, 1972. ix, 159 p.

Excellent presentation of chemical intrusion--natural and man-induced--in aquatic systems. Describes heavy metals and chlorinated hydrocarbon compounds. Many references.

227 Dyrssen, David, and Jagner, Daniel, eds. THE CHANGING CHEMISTRY OF THE OCEANS. Nobel Symposium 20. New York: John Wiley & Sons, 1972. 365 p.

Twenty-one contributors to an excellent survey volume; all major aspects of oceanography and marine pollution are included.

228 Edmondson, W.T. "Fresh Water Pollution." In ENVIRONMENT--RESOURCES, POLLUTION AND SOCIETY, edited by William W. Murdoch, pp. 251-72. 2d ed. Sunderland, Mass.: Sinauer Associates, 1975.

Excellent general treatment and referencing by a famous limnologist.

229 Epstein, Samuel S. "Control of Chemical Pollutants." NATURE 228 (1970): 816-19.

Major review of basic problems, methods of detecting and anticipating hazards, and mechanisms for more effective management. Makes many suggestions, including the formation of an interdisciplinary agency for maintenance of environmental quality and consumer safety.

230 Epstein, Samuel S., and Hattis, D. "Pollution and Human Health." In ENVIRONMENT--RESOURCES, POLLUTION AND SOCIETY, edited by William W. Murdoch, pp. 195-221. 2d ed. Sunderland, Mass.: Sinauer Associates, 1975.

Broad article but informative; teratogenicity, mutagenicity, and epidemiology of wide range of environmental pollutants.

231 Goldberg, E.D., and Bertine, K.K. "Marine Pollution." In ENVIRONMENT--RESOURCES, POLLUTION AND SOCIETY, edited by William W. Murdoch, pp. 273-95. 2d ed. Sunderland, Mass.: Sinauer Associates, 1975.

First provides broad treatment of the physics of marine environments, then focuses on specific classes of pollutants: petroleum, heavy metals (lead, mercury, cadmium), and chlorinated hydrocarbons (particularly DDT and PCBs). Good references.

232 Hart, C.W., Jr., and Fuller, Samuel L.H. POLLUTION ECOLOGY

OF FRESHWATER INVERTEBRATES. New York: Academic Press, 1974. xiv, 389 p.

Contains ten taxonomically arranged chapters by individual authorities. Chapters concern protozoans, sponges, flatworms, leeches, aquatic earthworms, bryozoans, crayfishes, clams and mussels, snails, and insects. Extremely rich in life history information; indicates the significance of each group as pollution indicators. Highly recommended.

233 Mellanby, Kenneth. THE BIOLOGY OF POLLUTION. Institute of Biology's Studies in Biology, no. 38. London: Edward Arnold, 1972. 60 p.

Brief, general accounts of all major forms of pollution. Well done by well-known British biologist.

234 Moriarty, Frank. "Pollutants and Foodchains." NEW SCIENTIST (England), 16 March 1972, pp. 594-96.

Challenges concept of pollutant residues' successive concentration along food chains. Amounts of residues retained probably depend more on differing rates of metabolism and excretion in organisms.

235 National Academy of Sciences. TOXICANTS OCCURRING NATURALLY IN FOODS. 2d ed. Washington, D.C.: Committee on Food Protection, National Research Council, 1973. vii, 624 p.

Eighteen expert contributors, members of the Committee on Food Protection, deal with aspects that concern food scientists. It is important for environmentalists to understand natural occurrences of metals, nitrates, animal and plant toxins, and other potentially dangerous substances in order to separate their effects from those of pollutants.

236 Nelson-Smith, A. OIL POLLUTION AND MARINE ECOLOGY. New York: Plenum Press, 1973. 260 p.

Semitechnical presentation in nine chapters. Even gives instructions on how to clean up oil-soaked sea birds.

237 Nicholson, H. Page. "The Needs for Water Quality Models on Agricultural Watersheds." JOURNAL OF ENVIRONMENTAL QUALITY 4 (1975): 21-23.

Recognition that agricultural practices contribute to pollution and deterioration of water quality. Suggests mathematical modeling and computer use to predict consequences of agronomic practices in entire basins or regions.

238 Oak Ridge National Laboratory. NSF-RANN TRACE CONTAMINANTS

DIRECTORY. Oak Ridge, Tenn: Toxic Materials Information Center, 1971--. Monthly.

Directory and abstracting journal begun in 1971 under the auspices of the National Science Foundation (NSF) and the Research Applied to National Needs (RANN) Trace Contaminants program. It provides effective communication throughout these agencies.

239 Olson, Theodore A., and Burgess, Frederick J., eds. POLLUTION AND MARINE ECOLOGY. New York: Interscience Publishers, 1967. xvi, 364 p.

Twenty contributed technical papers in varied aspects. In six sections: (1) "Man's Resources in the Marine Environment"; (2) "Dynamics of the Littoral Marine Community"; (3) "Ecological Systems"; (4) "Energy Transfer"; (5) "Interactions between the Biota and the Chemical-Physical Environment"; (6) "Parameters of Marine Pollution." Provides the sort of background reference one should use when interpreting specific instances of pollution.

240 Panel on Chemicals and Health. President's Science Advisory Committee. CHEMICALS AND HEALTH. Washington, D.C.: Science and Technology Policy Office, National Science Foundation, 1973. xi, 211 p.

Broad treatment of many problems with chemicals in foods and environment.

241 Parslow, J.L.F., et al. "Ingested Pollutants in Puffins and Their Eggs." BIRD STUDY (Oxford) 19 (1973): 18-33.

Puffins Fratercula artica have declined remarkably at breeding colonies. DDE and PCB residue loads are high in winter, markedly less in breeding birds. Possibly there is seasonal difference in size of liver. Concentrations of heavy metals higher in feathers than in body, except for cadmium and zinc. But cadmium levels (12.9, 22.3 ppm--dry weight) in livers are higher than other sea birds so far examined. Study highlights need for more work on marine contaminants.

242 PROCEEDINGS OF INTERNATIONAL CONFERENCE ON TRANSPORT OF PERSISTENT CHEMICALS IN AQUATIC ECOSYSTEMS. Conference held at Ottawa, Canada, 1-3 May 1974, sponsored by National Research Council of Canada. 334 p.

Many brief papers covering a broad range; much emphasis on mercury and other heavy metals and organochlorines including pesticides and PCBs. Has four sections: (1) "Water and Sediment Transport and Pollution Distribution"; (2) "Physical and Biological Exchanges between Aquatic Ecosystems Compartments";

(3) "Food Chain Transport of Pollutants"; and (4) "Ecosystem Simulation and Mathematical Modeling."

243 Seba, Douglas B., and Corcoran, E.F. "Surface Slicks as Concentrators of Pesticides in the Marine Environment." PESTICIDES MONITORING JOURNAL 3 (1969): 190-93.

Slicks effectively concentrate residues. High biological activity in vicinity of slicks assist further concentration. Notably DDT, DDE.

244 Shih, H.H. A LITERATURE SURVEY OF OCEAN POLLUTION. Washington, D.C.: Institute of Ocean Science and Engineering, Catholic University of America, 1971. 110 p. Available from National Technical Information Service (NTIS), Springfield, Va.

Chief concerns are oil spills, waste disposal, thermal discharges, and the like. But there is brief mention of biocides and rather good general description of marine food chains, in each case centered around a classification of literature.

245 Smith, J.E., ed. "TORREY CANYON" POLLUTION AND MARINE LIFE. Cambridge: At the University Press, 1968. Charts, plates, some colored.

Ten chapters on full range of effects following the sinking of this celebrated oil tanker.

246 Ui, Jun. "A Few Coastal Pollution Problems in Japan." In THE CHANGING CHEMISTRY OF THE OCEANS, edited by Daniel Dyrssen and David Jagner, pp. 171-76. New York: John Wiley & Sons, 1972.

Short descriptions of chemical effects: (1) methyl mercury poisoning in humans--"Minamata disease"; (2) cadmium in shellfish and algae--ill effects in humans of storage of industrial effluents in sea foods; and (3) PCB--residues generally higher in Japanese waters and sea life than in other parts of the world. Some of the chemicals discussed are potentially synergistic.

247 Vernberg, F. John, and Vernberg, Winona B., eds. POLLUTION AND PHYSIOLOGY OF MARINE ORGANISMS. New York: Academic Press, 1974. 506 p.

Twenty-three technical papers organized in five sections: heavy metals; pesticides and PCBs; oil and dispersants; general; and factor interaction--synergistic effects of pollutants and environmental parameters. Excellent reference; could be used in course work involving pollution effects.

248 Walker, Kenneth C. "The Role of Pesticides in Pollution Management." RESIDUE REVIEWS 34 (1971): 163-72.

Discusses and evaluates "environmental quality." Identifies controversy. Calls for public information and education.

249 Woodwell, G[eorge]. M. "Effects of Pollution on the Structure and Physiology of Ecosystems." SCIENCE 168 (1970): 429-33.

Broadly predictable ecosystemic patterns are discernible in the behavior of toxic substances in the biosphere. Many changes possible but article includes especially simplification of the plant and animal community, shifts in the ratio of gross production to total respiration, and loss of part or all of the inventory of nutrients. Extrapolations based on radionuclides, herbicides, and DDT distritutions and effects. Very thoughtful review.

TOXICOLOGY

Toxicology is the study of toxic substances. Ordinarily, the "environment" considered by toxicologists is the human body or, for reasons of testing, the bodies of laboratory animals. Any chemical which might produce an unwanted biological effect is a candidate for study by toxicologists. Not many toxicologists identify themselves as environmentalists; but it is on the biological response of foreign chemicals--xenobiotics--that environmental toxicology rests. Some general sources of information follow.

250 Ahnja, Sut, et al., eds. CHEMICAL ANALYSIS OF THE ENVIRONMENT AND OTHER MODERN TECHNIQUES. New York: Plenum Press, 1973. ix, 384 p.

Five editors of subsections with several contributed papers in each. Of particular interest here are trace metals, pesticides, and PCBs.

251 Allen, Stewart E., et al. CHEMICAL ANALYSIS OF ECOLOGICAL MATERIALS. New York: John Wiley & Sons, A Halstead Press Book, 1974. Printed in Great Britain. x, 565 p.

Eight major sections: soils; vegetation; waters; inorganic constituents; organic constituents; pollutants; instrumental techniques; data processes. Eight appendixes and twenty-three pages of references.

252 ANALYTICAL BIOCHEMISTRY. New York and London: Academic Press, 1960--. Irregular. Approx. 3,000 pages per year.

An international journal on analytical biochemistry.

253 Anderson, P.D., and Weber, L.J. "Toxic Response as a Quantitative Function of Body Size." TOXICOLOGY AND APPLIED PHARMACOLOGY 33 (1975): 471-83.

A good recent analysis of an old problem in toxicology. One would think that toxic response is related to weight, but this is not so. From the days of Bliss's formula a dose-response curve has been calculable. (See C.B. Bliss. "The Size Factor in the Action of Arsenic upon Silkworm Larvae." JOURNAL OF EXPERIMENTAL BIOLOGY 13 (1936): 95-110. The formula referred to is $Y = a + b \log M/W^h$. Essentially this formula expresses a linear function between survival time and the dose of sodium arsenate. Assuming the size factor - W^h- is unity, the dose is proportional to weight. Actually the exponent varies with age and sex.) These authors provide another refinement. The derived curve makes more predictive sense than an LD50 or LC50 alone which are mean values on a linear scale. Fishes of different sizes were used in this study; toxic exposure was to dieldrin, potassium pentachlorophenate, potassium cyanides, and chlorides of copper, zinc, and nickel.

254 Applegate, Vernon C., et al. TOXICITY OF 4,346 CHEMICALS TO LARVAL LAMPREYS AND FISHES. Special Scientific Reports--Fisheries, no. 207. Washington, D.C.: Fish and Wildlife Service, Bureau of Sports, Fisheries and Wildlife, 1957. 157 p.

Toxicity testing on sea lamprey larvae, rainbow trout, and bluegill sunfish. Toxicities not uniform in the three species. Useful tabulation.

255 Casarett, Louis J., and Doull, John. TOXICOLOGY: THE BASIC SCIENCE OF POISONS. New York: Macmillan Co., 1975. xiii, 768 p. Tables, graphs, chemical formulas.

Thirty-five contributors cover vast, detailed range. Presented in four units: (1) "General Principles of Toxicology"; (2) "Systemic Toxicology"; (3) "Toxic Agents"; and (4) "Applications of Toxicology." Includes carcinogens, teratogens, animal toxins, forensic toxicology, pesticides, and many more headings. Excellent sourcebook.

256 deSerres, Frederick J., and Sheridan, William, eds. THE EVALUATION OF CHEMICAL MUTAGENICITY DATA IN RELATION TO POPULATION RISK. Environmental Health Perspectives, issue 6. Washington, D.C.: Government Printing Office (U.S. Department of Health, Education, and Welfare), 1973. xi, 235 p.

Many contributors; six parts ranging from social meanings to recommendations.

257 DeWitt, James B. "Methodology for Determining Toxicity of Pesticides to Wild Vertebrates." JOURNAL OF APPLIED ECOLOGY 3 (1966): supplement, 275-78.

Effects are proportional to amount, rate, and manner of application. Persistence of basic chemical or compound determines manner of storage in animal tissue, subsequently to be passed on to other animals in food chains.

258 Environmental Mutagen Society, Council. "Environmental Mutagenic Hazards." SCIENCE 187 (1975): 503-14.

A complete review of the genetic and regulatory aspects of screening for mutation-causing chemicals in the environment. Prepared by an eminent group of scientists. Individual mutagens are not named. An area which must be understood, given present concerns of potential chemical effects.

259 Epstein, Samuel S. "Environmental Determinants of Human Cancer." CANCER RESEARCH 34 (1974): 2425-35.

Many carcinogens are described but Epstein strongly characterizes aldrin and dieldrin as causing cancer. Dieldrin is a common environmental pollutant.

260 Erney, D. Ronald. "Rapid Screening Method for Analysis of Chlorinated Pesticide and Polychlorinated Biphenyl Residues in Fish." JOURNAL OF THE ASSOCIATION OF OFFICIAL ANALYTICAL CHEMISTS 57 (1974): 576-79.

Describes economical method applicable to routine analysis of fish samples.

261 Fishbein, L., ed. PESTICIDES. Chromatography of Environmental Hazards, vol. 3. New York: Elsevier Scientific Publishing Co., 1975. 820 p.

Eighteen chapters by chemical grouping with detailed analytical methodologies and references.

262 Fishbein, L., et al. CHEMICAL MUTAGENS: ENVIRONMENTAL EFFECTS ON BIOLOGICAL SYSTEMS. New York: Academic Press, 1970. xi, 364 p.

Strongly genetic approach. Ends with chapter nine titled "Drugs, Food Additives, Pesticides, and Miscellaneous Mutagens."

263 Floyd, M., and Sommers, L.E. "Determination of Total Mercury in Soils and Sediments." JOURNAL OF ENVIRONMENTAL QUALITY 4 (1975): 323-25.

Describes a simple analytical method for determining total mercury. Method accurate within 1-3 percent.

264 Gallay, Wilfred, et al., comps. ENVIRONMENTAL POLLUTANTS, SELECTED ANALYTICAL METHODS. Scientific Committee on Problems of the Environment (SCOPE), no. 6. Ann Arbor, Mich.: Ann Arbor Science Publishers, 1975. 277 p.

Report of studies by the International Council of Scientific Unions, Scientific Committee on Problems of the Environment (SCOPE), Working Group on Methodology of Determination of Toxic Substances in the Environment. Concentrates on heavy metals, but also DDT, PCB, and other pollutants.

265 Garrison, A.W., et al. "Confirmation of Pesticide Residues by Mass Spectrometry and NMR Techniques." In FATE OF ORGANIC PESTICIDES IN THE AQUATIC ENVIRONMENT, edited by Robert F. Gould, pp. 26-54. Washington, D.C.: American Chemical Society, 1972.

Mass spectrometry, especially in tandem with gas chromatography, and NMR (nuclear magnetic resonance) can now be used for the detection of pesticide residues in environmental samples.

266 Haque, Rizwanul, and Biros, Francis J., eds. MASS SPECTROMETRY IN PESTICIDE CHEMISTRY. New York: Plenum Press, 1974. xi, 348 p.

Analytical methodologies using the two techniques for the major pesticide classes. Twenty-two contributors.

267 Hayes, Wayland J., Jr. "The 90-Dose LD50 and a Chronic Factor as Measures of Toxicity." TOXICOLOGY AND APPLIED PHARMACOLOGY 11 (1967): 327-35.

Proposed method to communicate results in which repeated doses given. Leads to quantitative analysis of long-term studies.

268 Hollaender, Alexander, ed. CHEMICAL MUTAGENS: PRINCIPLES AND METHODS FOR THEIR DETECTION. 2 vols. New York: Plenum Press, 1971. lxiii, 310 p.; lxii, 300 p.

Broad array of technical papers including mention of pesticides and other pollutants that are potential mutagens, and the modes of action of such chemicals.

269 Horwitz, William, ed. OFFICIAL METHODS OF ANALYSIS OF THE ASSOCIATION OF OFFICIAL ANALYTICAL CHEMISTS. 12th ed. Washington, D.C.: Association of Official Analytical Chemists, 1975. 1,094 p.

Standard reference for many years; updated at frequent intervals.

270 Janson, B., et al. "Identification by GS-MS of Phenolic Metabolites of PCB and p,p' DDE Isolated from Baltic Guillemot and Seal." AMBIO 4 (1975): 93-97.

Must analyze not only for chlorinated hydrocarbons but also their molar metabolites. Phenolic PCBs and hydroxylated metabolites of DDE found; those substances were previously found in wild-taken specimens. Different metabolic pathways in seals and guillemot (bird).

271 JOURNAL OF CHROMATOGRAPHIC SCIENCE. Niles, Ill.: Seaton T. Preston, 1963--. Monthly. Approx. 600 pages per year.

Technical journal involving the analytical composition of various substances.

272 JOURNAL OF CHROMATOGRAPHY. New York: Elsevier Scientific Publishing Co., 1958--. At least monthly. Approx. 5,000 pages per year.

Technical information associated with chromatography. Usually involves original research work not previously published in a generally accessible language in other periodicals.

273 JOURNAL OF GAS CHROMATOGRAPHY. See JOURNAL OF CHROMATOGRAPHIC SCIENCE.

274 Malaiyandi, M., and Barrette, J.P. "Determination of Mercury in Soils and Biological Matrices by the Vanadium Pentoxide Digestive Procedure." ARCHIVES OF ENVIRONMENTAL CONTAMINATION AND TOXICOLOGY 2 (1974): 315-30.

Contrasts with use of potassium permanganate-sulfuric acid oxidant system; values are less reagent and shorter time.

275 Matsumura, Fumio. TOXICOLOGY OF INSECTICIDES. New York: Plenum Press, 1975. xvii, 503 p. Graphs, tables.

A toxicology text written by an entomologist rather than a clinician. This book will become a standard reference. Excellent coverage.

276 Mulhern, Bernard M., et al. "Semiquantitative Determination of Polychlorinated Biphenyls in Tissue Samples by Thin Layer Chromatography." JOURNAL OF THE ASSOCIATION OF OFFICIAL ANALYTICAL CHEMISTS 54 (1971): 548-50.

Necessary to convert DDE to DCBP; otherwise no prior separation of PCB from chlorinated pesticides required.

277 Peters, J.A., and Baxter, K.J. "Analytical Determination of Compound 1080 (Sodium Fluoroacetate) Residues in Biological Materials." BULLETIN OF ENVIRONMENTAL CONTAMINATION AND TOXICOLOGY 11 (1974): 177-83.

Very important report. Finally, a rapid and accurate method for assessing 1080 and metabolites in animal tissues. Chemical was used in vertebrate pest control, in this instance, in New Zealand.

278 Prakash, Louise, et al., eds. MOLECULAR AND ENVIRONMENTAL ASPECTS OF MUTAGENESIS. Springfield, Ill.: Charles C. Thomas, Publisher, 1974. iv, 289 p.

Individual authorship for sixteen contributions, which are divided into five reporting sessions. Good range of coverage.

Proceedings of the Rochester (N.Y.) International Conferences on Environmental Toxicity, sixth conference, 4-6 June 1973.

279 Reynolds, L.M. "Pesticide Residue Analysis in the Presence of Polychlorobiphenyls (PCBs)." RESIDUE REVIEWS 34 (1971): 27-57.

Excellent review of all phases of interest in PCBs, particularly interference on chromatographs. Good literature list. Summarizes organochlorine and PCB residues in Canadian wildlife.

280 Richardson, A., et al. "Residues of Polychlorobiphenyls in Biological Samples." PESTICIDE MONITORING JOURNAL 4 (1971): 169-76.

PCBs in samples from aquatic origins are confirmed. Analytical difficulties and comparison with insecticides described.

281 SEPARATION SCIENCE. New York: Marcel Dekker, 1966--. Bimonthly. Approx. 850 pages per year.

An interdisciplinary journal of methods and underlying processes. Scientific articles; technical.

282 Shepard, Thomas H. CATALOG OF TERATOGENIC AGENTS. Baltimore, Md.: Johns Hopkins University Press, 1973. xvii, 211 p.

A catalog, listed by chemical name; attributes and references are given.

283 Stevens, Alan A., and Robertson, Eric A. "Determination of Methylmercury in Water." ARCHIVES OF ENVIRONMENTAL CONTAMINATION AND TOXICOLOGY 2 (1974): 266-74.

Method for routine identification and quantification of individual compounds in low concentrations in water. Good recovery range of 0.37 to 37 ppb. Method not subject to interference from mercuric ion.

284 Sunshine, Irving, ed. HANDBOOK OF ANALYTICAL TOXICOLOGY. Cleveland, Ohio: CRC Press, 1969. 1,081 p.

Thirty-eight advisors and contributors; tends to be chemically oriented but does include biological aspects of all major chemical groups.

285 Sutton, H. Eldon, and Harris, Maureen I., eds. MUTAGENIC EFFECTS OF ENVIRONMENTAL CONTAMINANTS. Environmental Sciences Monograph series. Douglas A.K. Lee et al., general eds. New York: Academic Press, 1972. xiv, 195 p.

Separate contributors to fourteen sections beginning with "Genetic Toxicology" and ending with "Possible Relationships between Mutagenesis and Carcinogenesis."

286 Teeny, Faud M. "Rapid Method for the Determination of Mercury in Fish Tissue by Atomic Absorption Spectroscopy." JOURNAL OF AGRICULTURAL AND FOOD CHEMISTRY 23 (1975): 660-71.

Simple and rapid method; digestion time fifteen minutes. Results compare to Food and Drug Administration "Official First Action Method."

287 Thompson, William R. "Use of Moving Averages and Interpolation to Estimate Median Effective Dose. I. Fundamental Formulas, Estimation of Error, and Relation to Other Methods." BACTERIOLOGICAL REVIEWS 11 (1947): 115-45.

Biological assay gives action and potency of reagent. Tests designs and methods. Choice of test influences conclusion and judgment.

288 U.S. Department of Health, Education, and Welfare. THE TOXIC SUBSTANCES LIST, 1974 EDITION. HEW Publication, no. (NIOSH) 74-134. Washington, D.C.: 1974. xciii, 904 p.

Source book of chemicals judged to be toxic in some human relationship. Includes 13,000 different chemicals in 42,000 different formulations. Chemical names, applicable regulations, references, cross-referencing. Probably will become an annual publication.

289 Weast, Robert C., ed. HANDBOOK OF CHEMISTRY AND PHYSICS. 50th ed. Cleveland, Ohio: Chemical Rubber Co., 1970. xxiv, Sections A-F, separately numbered.

The major summation of technical data points in chemistry and physics.

290 Weil, Carrol S. "Tables for Convenient Calculation of Median Effective Dose (LD50 or ED50) and Instructions in Their Use." BIOMETRICS 8 (1952): 244-63.

Reduces time of calculation of median effective dose and its confidence limits.

Part II

CONSEQUENCES OF INTENTIONAL ENVIRONMENTAL POLLUTION

Section 4

PESTICIDES

Any chemical introduced into a biological environment with the intention of killing organisms or of altering biological functioning can be called a pesticide. Most chemicals are directed against insects; fewer are directed against plants; and fewer yet against vertebrate animals. Many chemical types and biological reactions are exploited. The environmental toxicologist is concerned with all such chemicals and does the testing to determine hazard in their use. The evaluation of effects on nontarget species is an important aspect of the environmentalist's approach to the effects of pollutant chemicals. The following citations illustrate the broad range of chemicals, target species, and biological consequences involved and with which environmentalists are concerned.

291 Adams, Russell S., Jr. "Factors Influencing Soil Adsorption and Bioactivity of Pesticides." RESIDUE REVIEWS 47 (1973): 1-54.

Important technical review and bibliography. Soils affect pesticide behavior. Important factors: clay content and particle size, surface area, moisture, and most important, organic material. Regulations setting tolerances or thresholds must take highly differing environmental conditions into account.

292 American Chemical Society. "Decontamination of Pesticide Residues in the Environment." RESIDUE REVIEWS 29 (1969): entire volume. 213 p.

The "environment" described in most of the thirteen papers in the symposium is crop or livestock production or food processing. For other approaches see particularly P.C. Kearney et al. on soils in this issue, and R.B. Hemmett, Jr., and S.D. Faust, on 2,4-D breakdown by aquatic microorganisms, in the same issue.

Report based on Atlantic City meetings of the American Chemical Society in September 1968.

293 Aulerich, R.J., et al. "Effects of Feeding Coho Salmon and Other Great Lakes Fish on Mink Reproduction." CANADIAN JOURNAL OF ZOOLOGY 49 (1971): 611-16.

Mink reproduction showed no correlation with oxidative rancidity or mercury levels in fish, but seemingly did correlate with pesticide residues. Reproductive decline and kit mortality related to residue levels in tissues. Residue levels depend on type of fish and the microenvironment from which the type comes.

294 Bartsch, Elisabeth, et al. "The Carbinole Acaricides: Chlorobenzilate and Chloropropylate." RESIDUE REVIEWS 39 (1971): entire volume. 88 p.

Carbinole acaricides were developed originally for spider mites and became significant when mites developed resistance to organophosphorus compounds. Very low acute oral and dermal toxicity to mammals; lower than DDT. Quickly translocated into plants. Tolerance level 5.0 ppm. Article includes long bibliography.

295 Boddicker, M.L., et al. "Parasites and Pesticide Residues of Mountain Goats in South Dakota." JOURNAL OF WILDLIFE MANAGEMENT 35 (1971): 94-103.

Fat samples from mountain goats *Oreamnos* from the Black Hills averaged 0.59 ppm of heptachlor epoxide. Samples also contained dieldrin, DDE, and DDD. Goats sampled were remote from agricultural areas. Amounts from goats are more than triple amounts in deer and are six times that in antelope from same area.

296 Brown, Royal L. PESTICIDES IN CLINICAL PRACTICE. Springfield, Ill.: Charles C. Thomas, Publisher, 1966. xi, 492 p.

Identification, pharmacology, and therapeutics of pesticides. Encyclopedic style. Lists 8,532 pesticide formulations by brand name. Useful source.

297 Carson, Rachel. SILENT SPRING. Boston: Houghton Mifflin Co., 1962. ix, 368 p.

The classic and controversial book that first widely announced the hazards of pesticides. Has appeared subsequently in many printings and translations. Fine literary style, and scientifically accurate.

298 Casey, James E., et al. "An Example of Economically Feasible Opportunities for Reducing Pesticide Use in Commercial Agriculture." JOURNAL OF ENVIRONMENTAL QUALITY 4 (1975): 60-64.

A practical, economically rewarding, pest management strategy to reduce insecticide use in control of cotton insects. Based on lower volumes of use, more widely spaced applications, and time to allow populations of insects to increase. Costs were down and yields were up in this southern Texas study.

299 Chichester, C.O., ed. RESEARCH IN PESTICIDES. New York: Academic Press, 1965. xv, 380 p.

Unusually well-balanced presentation by thirty-nine contributors. Sections include public policy, pesticide actions, whole animal responses, ecological systems, removal and/or persistence, alternate methods of pest control, and relation to human health.

300 Cohen, Jesse M., and Pinkerton, Cecil. "Widespread Translocations of Pesticides by Air Transport and Rain-out." In ORGANIC PESTICIDES IN THE ENVIRONMENT, edited by Aaron A. Rosen and H.F. Kraybill, pp. 163-76. Advances in Chemistry Series, no. 60. Washington, D.C.: American Chemical Society, 1966.

Pesticides have intruded into areas of the world not treated in any way. Usually food or water are means of translocation. Atmosphere also provides a route. Rainfall and dust samples were examined; all had some organochlorine compounds in them.

301 Crockett, A.B., et al. "Pesticide and Mercury Residues in Commercially Grown Catfish." PESTICIDES MONITORING JOURNAL 8 (1975): 235-40.

Pesticide residues in 15 percent of samples were above FDA action levels. Related to treatment of adjoining soybean fields, not to water or food. Probable source of residue is treatment of cotton. Aerial transport seems most likely route of translocation.

302 Croft, W.A., and Brown, A.W.A. "Responses of Arthropod Natural Enemies to Insecticides." ANNUAL REVIEW OF ENTOMOLOGY 20 (1975): 285-335.

In the last quarter-century the study of insecticides and their physical and biological side effects has developed into a precise science. This summary is an excellent historical and present overview of this science. Maximum use of natural enemies of pests is now encouraged. Direct competitors are emphasized. The report covers literature to December 1973. Literature lists includes 246 citations, selected from topical and international sources.

303 Crosby, Donald G. "The Intentional Removal of Pesticide Residues." In RESEARCH IN PESTICIDES, edited by C.O. Chichester, pp. 213-23. New York: Academic Press, 1965.

Thoughtful set of questions. Just how do we logically set out to remove environmental residues?

304 DaSilva, Edgar J., et al. "Effect of Pesticides on Blue-Green Algae and Nitrogen-Fixation." ARCHIVES OF ENVIRONMENTAL CONTAMINATION AND TOXICOLOGY 3 (1975): 193-204.

Reports on six herbicides, one organophosphate insecticide (malathion), nine nitrogen-fixing algae. Concludes that some pesticides severely limit nitrogen fixation, thereby affecting overall nitrogen economy in soils.

305 Davies, John E., et al. "Problems of Prevalence of Pesticide Residues in Humans." PESTICIDES MONITORING JOURNAL 2 (1968): 80-85.

Greater number and variety of samples are needed before demographic comparisons are valid. Whole blood and fat are equally reliable and blood is easier to get.

306 Durham, William F., and Williams, C.H. "Mutagenic, Teratogenic, and Carcinogenic Properties of Pesticides." ANNUAL REVIEW OF ENTOMOLOGY 17 (1972): 123-48.

Inadequacies at present in interpretation of animal data. Some carcinogens, but human data meager. The current Community Studies Programs of the Environmental Protection Agency may help in time. Low-level residue intake studies particularly needed. Eighty-seven references.

307 Edwards, Clive A., ed. ENVIRONMENTAL POLLUTION BY PESTICIDES. London and New York: Plenum Press, 1973. xiii, 542 p.

Expanded version of Edward's 1970 PERSISTENT PESTICIDES IN THE ENVIRONMENT, employing individual specialists as contributors. Chapter 1 deals with analytical methods for determining pesticide residues; the remaining twelve contributions describe residue problems in many living forms, from plants to man. Technically oriented; many references.

308 Edwards, C[live]. A., and Thompson, A.R. "Pesticides and the Soil Fauna." RESIDUE REVIEWS 45 (1973): 2-79.

Extensive summary and bibliography on little understood segment of the biosphere. All groups of chemicals and soil animals considered. Major effects on nontarget organisms are (1) saprophagous organisms--those that break down organic matter; (2) predators--those that maintain population balances. Chlorinated hydrocarbons and some herbicides (e.g., monuron) present the chief difficulties. Persistent chemicals accumulate in invertebrate tissues. Normally innocuous species may become pests because of chemically induced population imbalances.

309 Epstein, Samuel S., and Legator, Marvin S., eds. MUTAGENICITY OF PESTICIDES: CONCEPTS AND EVALUATION. Cambridge: Massachusetts Institute of Technology Press, 1971. xvii, 220 p.

Importance of, testing for, and literature on mutagenicity. Many contributors.

310 Flickinger, Edward L., and Meeker, Dennis L. "Pesticide Mortality of Young White-faced Ibis in Texas." BULLETIN OF ENVIRONMENTAL CONTAMINATION AND TOXICOLOGY 8 (1972): 165-68.

Dead young ibis had residues of aldrin, DDE, dieldrin, and mercury. Latter two were at sufficiently high concentrations to presume causing death. Residues are present in invertebrate animals from treated rice fields. Parents feed these invertebrates to their young.

311 Foster, Thomas S. "Physiological and Biological Effects of Pesticide Residues in Poultry." RESIDUE REVIEWS 51 (1974): 70-121.

Major review and bibliography on poultry that should be studied as well as wild species of birds. Most chemicals are not acutely harmful. Some chemicals or their residues are. Most organophosphates are teratogenic. Some chemicals, for example, ethylene bromide, and the fungicides arasan, thiram, zineb, and ziram, cause decrease in egg size. Organochlorine residues do not cause eggshell thinning when 3 percent calcium is added to diet. Several organochlorines decrease fertility and hatchability, and several affect hepatic microsomal enzymes. Kepone and other chlorinated polycyclic compounds have androgenic effects.

312 Gaines, Thomas B. "The Acute Toxicity of Pesticides to Rats." TOXICOLOGY AND APPLIED PHARMACOLOGY 2 (1960): 88-99.

Study of relative acute toxicity of chemicals based on LD50 determinations. Gaines reports LD50 for forty-two pesticides and two metabolites of DDT administered in single doses by oral or dermal route to adult Sherman-strain rats.

313 Gerakis, P.A., and Sficas, A.G. "The Presence and Cycling of Pesticides in the Ecosphere." RESIDUE REVIEWS 52 (1974): 69-87.

Excellent broad coverage, with long bibliography, of the fates of applied chemicals in relation to crop production throughout the world.

314 Gould, Robert F., ed. FATE OF ORGANIC PESTICIDES IN THE AQUATIC ENVIRONMENT. Advances in Chemistry Series, no 111. Washington, D.C.: American Chemical Society, 1972. viii, 280 p.

Symposium papers from a meeting of the American Chemical

Society, Los Angeles, California, 29-31 March 1971. Thirteen contributors covering (1) occurrence, recovery, and identification; (2) distribution and transport mechanisms; (3) stability and metabolites from chemical systems; and (4) stability and metabolites from biological systems.

315 Greenberg, Robert E., and Heye, Paul L. "Insecticide Residues in Little Blue Herons." WILSON BULLETIN 83 (1971): 95-97.

Study of herons (Florida caerulea) from Missouri. Eggs, nestlings, and adults analyzed for DDT, aldrin-dieldrin, and heptachlor. More DDT residues found than other residues; p,p'-DDE found in all. Maximum p,p'-DDE in eggs, 2.78 ppm; in muscle, 5.56 ppm; in brain of adults, 2.09 ppm. Table on all values. For adult herons, residues are from foods; for eggs and nestlings, residues come from stored amounts in adults. No change in population numbers or nesting success.

316 Haque, R[izwanul]., and Freed, V.H. "Behavior of Pesticides in the Environment: 'Environmental Chemodynamics'." RESIDUE REVIEWS 52 (1974): 89-116.

Chemodynamics is relationship between physico-chemical principles and chemical behavior in the environment. Major review with extensive bibliography. Covers behavior of pesticides in air, water, soils, and biota. Provides transport and degradation rates.

317 ______, eds. ENVIRONMENTAL DYNAMICS OF PESTICIDES. Environment Science Research Series, no. 6. New York: Plenum Press, 1975. 397 p.

Proceedings of a 1974 symposium held in Los Angeles by the Division of Pesticide Chemistry, American Chemical Society. Includes contributed papers and bibliographies.

318 Herman, Steven G., et al. "Pesticides and the Western Grebe." In CHEMICAL FALL-OUT, edited by M.W. Miller and G.G. Berg, pp. 24-53. Springfield, Ill.: Charles C. Thomas, Publisher, 1969.

Illustrates classic, well-studied case of reproductive inhibition and mortality in birds due to accumulated pesticide residues (DDD and DDE, chiefly).

319 Hill, Elwood F. "Avoidance of Lethal Dietary Concentrations of Insecticides by House Sparrows." JOURNAL OF WILDLIFE MANAGEMENT 36 (1972): 635-39.

House sparrows can detect pesticidal chemicals (abate, bromophos, gardona, and DDT) in food and will normally avoid if given an option. They turn to contaminated food only when starving.

320 _____. "Toxicity of Selected Mosquito Larvicides to Some Common Avian Species." JOURNAL OF WILDLIFE MANAGEMENT 35 (1971): 757-62.

Emphasizes LC50--the lethal concentration in diet for 50 percent of the test population. Important because LD50 indicates acute toxicity, not a likely thing in field situations (although toxicologists still depend on LD values in laboratory species for making human and environmental judgments regarding safety margins). Birds were bobwhite, blue jay, house sparrows, and cardinals; chemicals studied, in order of decreasing toxicity, were abate, DDT, bromophos, and gardona. Responses are directly related to size except for jays which are surprisingly sensitive.

321 Hurlbert, Stuart H. "Secondary Effects of Pesticides on Aquatic Ecosystems." RESIDUE REVIEWS 57 (1975): 82-148.

Excellent compilation and references. Tables of toxicity.

322 Hurtig, Henry. "Long-Distance Transport of Pesticides." In ENVIRONMENTAL TOXICOLOGY OF PESTICIDES, edited by F. Matsumura et al., pp. 257-80. New York: Academic Press, 1972.

Useful viewpoints on means of transport over great distances by persistent residues.

323 Kearney, Philip C., et al. "Decontamination of Pesticides in Soils." RESIDUE REVIEWS 29 (1969): 137-49.

Describes variously induced pathways of degradation under experimental conditions.

324 Kenaga, Eugene E. "Factors Related to Bioconcentration of Pesticides." In ENVIRONMENTAL TOXICOLOGY OF PESTICIDES, edited by F. Matsumura et al., pp. 193-228. New York: Academic Press, 1972.

Fine analysis of concentration factors and residue partitioning.

325 _____. "Guidelines for Environmental Study of Pesticides: Determination of Bioconcentration Potential." RESIDUE REVIEWS 44 (1972): 73-113.

Fine, extensive summation; long bibliography. Calls for standardization of environmentally related test methods. Some factors leading to high bioaccumulation are low water solubility; high fat solubility; high partitioning coefficients comparing water to its living components; and high stability under hydrolytic, light, heat, and microbiological conditions.

326 Khan, M.A.Q., and Bederka, John P., Jr., eds. SURVIVAL IN

TOXIC ENVIRONMENTS. New York: Academic Press, 1974. xxi, 553 p.

Derived from a symposium titled "Mechanisms of Survival in Toxic Environments," 28-30 December 1973, American Society of Zoologists. Twenty-four contributions in five sections: (1) "Impact of Chemical Pollutants on the Biology of Organisms"; (2) "Detoxication Mechanisms of Survival in Toxic Environments"; (3) "Role of Mixed-Function Oxidase and Its Components in Survival in Toxic Environments"; (4) "Recent Trends in Pesticide Research"; (5) "Non-Pesticidal Pollutants." Excellent technical contributions.

327 Kilgore, Wendell W., and Doutt, Richard L., eds. PEST CONTROL--BIOLOGICAL, PHYSICAL, AND SELECTED CHEMICAL METHODS. New York: Academic Press, 1967. xii, 477 p.

Very useful survey in eleven chapters by excellent authors. Chapter titles identify coverage: (1) "Biological Control"; (2) "Microbial Pesticides"; (3) "Electromagnetic Energy"; (4) "Radiation-Induced Sterilization"; (5) "Chemosterilants"; (6) "Pheromones"; (7) "Repellents"; (8) "Antifeedants"; (9) "Integrated Control"; (10) "Biocontrol and Chemosterilants [vertebrates]"; (11) "Behavioral Manipulation (Visual, Mechanical, and Acoustical [vertebrates])."

Rare to find such a range of topics.

328 Kilgore, W[endell]. W., and Li, Ming-Yu. "The Carcinogenicity of Pesticides." RESIDUE REVIEWS 48 (1974): 141-61.

Of the greatest concern to man. Some pesticides are tumorigenic, a few could be carcinogenic. Little known on photodecomposition products as they affect man. Inconclusive relationship generally.

329 Klemmer, Howard W. "Human Health and Pesticides--Community Pesticide Studies." RESIDUE REVIEWS 41 (1972): 55-63.

Describes survey methods in thirteen states; sponsored by Environmental Protection Agency. Longer studies in Hawaii described. Positive correlations shown between residue levels and human symptoms (sinus trouble, bronchitis, and asthma).

330 Kreitzer, J.F., and Heinz, Gary H. "The Effect of Sublethal Dosages of Five Pesticides and a Polychlorinated Biphenyl on the Avoidance Response of Coturnix Quail Chicks." ENVIRONMENTAL POLLUTION 6 (1974): 21-29.

Chicks given in diet: chlordane, dieldrin, endrin, DDE, ceresan M (10.4 percent mercury compounds), or Arochlor 1254 (PCB). Group exhibited avoidance response to all chemicals but DDE. Total recovery of normal behavioral responses occurred upon cessation of treatment in all insecticides, but not with mercury and PCB. Implication is that any chemical residue suppressing avoidance responses would affect survivability in the field.

331 Kreitzer, J.F., and Spann, J.W. "Tests of Pesticidal Synergism with Young Pheasants and Japanese Quail." BULLETIN OF ENVIRONMENTAL CONTAMINATION AND TOXICOLOGY 9 (1973): 250-56.

Wild species frequently have several different pollutant residues in tissue. It is not known whether effects are additive or one enhances the toxicity of another (synergistic). In this study of eleven pair combinations of chemicals only one pair was synergistic. These chemicals were malathion and trichlorofon when fed together. Probably such synergism is not especially severe.

332 Librarian, Agriculture Canada. PESTICIDE APPLICATION AND SPRAYING: A SELECTED BIBLIOGRAPHY. Saskatoon, Sask.: Research Branch, Agriculture Canada, 1974.

Not annotated or paginated; divided into fifteen headings from aerial application to ultralow volume application.

333 Lichtenstein, E. Paul, and Schulz, Kenneth R. "Volatilization of Insecticides from Various Substrates." JOURNAL OF AGRICULTURAL AND FOOD CHEMISTRY 18 (1970): 814-18.

Comparisons of rates in four chlorinated hydrocarbon and four organophosphorus insecticides. Aldrin most volatile, azinphosmethyl least. Highest volatilization rates from water, least from soils.

334 Malone, C[harles]. R., and Blaylock, B.G. "Toxicity of Insecticide Formulations to Carp Embryos Reared in Vitro." JOURNAL OF WILDLIFE MANAGEMENT 34 (1970): 460-63.

None of the chemicals (DDT, chlordane, dieldrin, endrin, diazinon, and guthion) affected viability at less than 1.0 ppm, but all did at 5.0 ppm (50-100 percent mortality).

335 Matsumura, Fumio. "Microbial Degradation of Pesticides." In SURVIVAL IN TOXIC ENVIRONMENTS, edited by M.A.Q. Khan and John B. Bederka, Jr., pp. 129-54. New York: Academic Press, 1974.

Pesticides are viewed as environmental microcontaminants which, although in small quantities, have the capacity to make major effects on biological systems. Detailed review emphasizing biological aspects. Long reference list.

336 Matsumura, Fumio, et al., eds. ENVIRONMENTAL TOXICOLOGY OF PESTICIDES. New York: Academic Press, 1972. xiv, 637 p.

Twenty-nine contributions by specialists. Proceedings of a U.S.-Japan seminar at Oiso, Japan, October 1971. Eight parts: patterns of pesticide usage and residue occurrence; mercury transformations; chlorinated hydrocarbons; fungicides, herbicides, organophosphates, and carbamates; microbial influences on degradation; photodecomposition of pesticides; toxic effects of residues on wildlife; design of new pesticides (pheromones, fungicides).

337 Melnikov, N.N. "Chemistry of Pesticides." Translated by Ruth L. Busbey. RESIDUE REVIEWS 36 (1971): entire volume. x, 480 p.

Translation of Russian comprehensive work on chemicals used in pest control. Updated version of 1968 edition. Introduction to extensive Russian literature.

338 Menzie, C[alvin]. M. "Fate of Pesticides in the Environment." ANNUAL REVIEW OF ENTOMOLOGY 17 (1972): 199-222.

The chemistry of degradation, chiefly in organochlorines and organophosphorus compounds. Contains 148 references.

339 ______. METABOLISM OF PESTICIDES. Special Scientific Report--Wildlife, no. 127. Washington, D.C.: Fish and Wildlife Service, Bureau of Sports Fisheries and Wildlife, 1969. 487 p.

"Disappearance" mechanisms of pesticides by metabolic means. Chemical sequences given alphabetically under common name of pesticide. Contains 1,792 references.

340 Metcalf, Robert L., et al. "Model Ecosystem for the Evaluation of Pesticide Biodegradability and Ecological Magnification." ENVIRONMENTAL SCIENCE AND TECHNOLOGY 5 (1971): 709-13.

Small laboratory model systems with a terrestrial-aquatic interface and a seven-element food chain. Simulates pesticide applications to crop plants and eventual contamination to aquatic environments. In this study evaluations were made with radiolabeled DDT, DDD, DDE, and methoxychlor. DDE, DDT, DDD concentrated 10,000- to 100,000-fold in system. DDE concentrated 30,000-fold and was stored with little metabolism. Methoxychlor was readily metabolized and was stored at comparatively low levels. This model can be used for industrial chemicals as well. This laboratory method of testing concentration abilities in controlled ecosystems is important in pesticide testing for EPA registration.

341 Miller, Morton W., and Berg, George G., eds. CHEMICAL FALLOUT:

CURRENT RESEARCH ON PERSISTENT PESTICIDES. Springfield, Ill.: Charles C. Thomas, Publisher, 1969. xxii, 531 p.

A proceedings publication of the University of Rochester conferences on toxicity. Twenty-four excellent contributions centering on effects of organochlorine and organomercurial compounds. Soils, fishes, birds, and human beings are discussed. Printed questions and answers following papers are illuminating.

342 Moore, Norman W. "A Synopsis of the Pesticide Problem." ADVANCES IN ECOLOGICAL RESEARCH 4 (1967): 75-129.

Extensive review of pest control and pesticide effects, viewed by a wildlife specialist in Great Britain.

343 , ed. "Pesticides in the Environment and Their Effects on Wildlife." JOURNAL OF APPLIED ECOLOGY 3 (1966): supplement. xii, 311 p.

The proceedings of an Advanced Study Institute sponsored by the North Atlantic Treaty Organization (NATO) at Monks Wood Experimental Station, England, 1-14 July 1965. Twenty-seven contributors on the effects of pesticides on various animals. American and European environments. Excellent references.

344 Moriarty, F[rank]. "The Effects of Pesticides on Wildlife: Exposure and Residues." SCIENCE OF THE TOTAL ENVIRONMENT (Amsterdam) 1 (1972): 267-88.

Worth reading; the "heresy" of some ideas, such as trophic concentration, is more apparent than real. Predators select contaminated prey; they do not build up residues slowly in themselves.

345 Muirhead-Thompson, R.C. PESTICIDES AND FRESHWATER FAUNA. London and New York: Academic Press, 1971. vii, 248 p.

Fine coverage and reference list. The only book with an international outlook on such coverage. Both intentional and unintended results discussed.

346 Paris, Doris F., and Lewis, David L. "Chemical and Microbial Degradation of Ten Pesticides in Aquatic Systems." RESIDUE REVIEWS 45 (1973): 95-124.

Literature search (very long bibliography) on chemical and microbial degradation for these chemicals: atrazine, captan, carbaryl, diazinon, 2,4-D, malathion, methoxychlor, parathion, PCBs, and toxaphene. Survey reveals there are no reported degradation products for PCBs and toxaphene. Photochemical products for only seven of ten chemicals are known. Only 2,4-D is studied in detail regarding microbial degradation and measurements thereof.

347 Peakall, David B. "Pesticides and the Reproduction of Birds." SCIENTIFIC AMERICAN 222 (1970): 72-78.

Semipopular review of the various mechanisms and physiological theories by which reproductive processes in birds might be inhibited by toxic synthetic chemical residues in tissues. Relates chiefly to DDT metabolites and calcium metabolism.

348 Phillips, John H. "Effect of Pesticide Chemicals on Marine Organisms." In AGRICULTURAL CHEMICALS--HARMONY OR DISCORD FOR FOOD, PEOPLE AND THE ENVIRONMENT, edited by J.E. Swift, pp. 63-79. 1971.

Proceedings of a symposium sponsored by the Division of Agricultural Sciences, University of California, 16-17 February 1971. Although the paper itself is a brief, data-filled review, the report of the question and answer period informs more about the range of problems and attitudes than does the paper.

349 Pimentel, David. ECOLOGICAL EFFECTS OF PESTICIDES ON NON-TARGET SPECIES. Executive Office of the President, Office of Science and Technology, GPO no. 4106-0029. Washington, D.C.: Government Printing Office, 1971. 220 p.

Useful compilation and commentary with many references.

350 Pimentel, David, and Goodman, Nancy. "Environmental Impact of Pesticides." In SURVIVAL IN TOXIC ENVIRONMENTS, edited by M.A.Q. Khan and John B. Bederka, Jr., pp. 25-52. New York: Academic Press, 1974.

Some pesticides have done the following: (1) influenced structure and functions of ecosystems; (2) reduced species population numbers in some regions; (3) altered natural habitat; (4) stimulated or suppressed growth in some biota; (5) increased or decreased reproductive capacity of animals; (6) altered nutritional content of food; (7) increased susceptibility of certain plants and animals to disease and predation; (8) changed natural evolution of species populations in some regions. Excellent review and references.

351 Rosen, Aaron A., and Kraybill, H.F., eds. ORGANIC PESTICIDES IN THE ENVIRONMENT. Advances in Chemistry Series, no. 60. Washington, D.C.: American Chemical Society, 1966. xi, 309 p.

Twenty-three papers on many aspects of pesticide use, chiefly methodological but including several environmental studies.

352 Rosenberg, D.M. "Food Chain Concentration of Pesticides in Inverte-

brate Communities: A Re-evaluation." QUAESTIONES ENTOMOLOGICAE (Edmonton, Alta.) 11 (1975): 97-110.

Important challenge to concept of food chain or trophic level concentration of chlorinated hydrocarbon residues in animal tissues. Emphasizes his own work with dieldrin in freshwater pond invertebrates, but extends beyond. Suggests that habitat and mode of life rather than trophic level are most important ecological characteristics determining uptake and final concentrations. Appropriate term should be "bioconcentration." Actually this word and "biological magnification" are more commonly used; but food-chain concept is still valuable.

353 Rudd, R[obert]. L. "Pesticides." In ENVIRONMENT--RESOURCES, POLLUTION AND SOCIETY, edited by W.W. Murdoch, pp. 325-53. 2d ed. Sunderland, Mass.: Sinauer Associates, 1975.

A summary view of practical, toxicological, and ecological matters concerned with pesticides. Readable account; extensive reference list.

354 _____. PESTICIDES AND THE LIVING LANDSCAPE. Madison: University of Wisconsin Press, 1964. xi, 320 p.

A general scholarly study of the influence of pesticides in nature. Twenty-two chapters. Same text published in Canada and in England (Faber and Faber, 1966). Later translation into Swedish appeared under the title BIOCIDER OCH DEN LEVANDE NATUREN (Stockholm: Wahlstrom-Widstrand, 1968). Also cited in section 8: Vertebrate Pest Control.

355 Rudd, R[obert]. L., and Genelly, R.E. PESTICIDES: THEIR USE AND TOXICITY IN RELATION TO WILDLIFE. California Department of Fish and Game, Game Bulletin no. 7. Sacramento: State Printing Office, 1956. 209 p.

The first extensive review of all classes of pesticides as they affect plants, animals, and environment. Also cited in section 8: Vertebrate Pest Control.

356 Rudd, Robert L., and Herman, Steven G. "Ecosystemic Transferral of Pesticide Residues in an Aquatic Environment." In ENVIRONMENTAL TOXICOLOGY OF PESTICIDES, edited by Fumio Matsumura et al., pp. 471-85. New York: Academic Press, 1972.

DDD and DDE residues in Clear Lake, California, are found throughout the entire aquatic ecosystem. Ultimately in a fish-eating bird (Western grebe) accumulated residues were suffiiently high to cause mortality and reproductive inhibition. In 1968 a diminution in egg residues "allowed" greater survival of young grebes. This is attributed to fish introduced in 1967, increasing fish biomass and creating a "nutrient shunt." Fewer residues were available in the food of grebes.

357 Sanders, H.O., and Cope, Oliver B. "Toxicities of Several Pesticides to Two Species of Cladocerans." TRANSACTIONS OF THE AMERICAN FISHERIES SOCIETY 95 (1966): 165-69.

Wide array of pesticides tested against these sensitive aquatic organisms.

358 Sethunathan, N. "Microbial Degradation of Insecticides in Flooded Soil and in Anaerobic Cultures." RESIDUE REVIEWS 47 (1973): 143-65.

Rice is the staple food of half the world; hence the importance of understanding pesticide removal. Rapidly degraded chemicals in flooded soils are lindane, BHC, heptachlor, DDT, methoxychlor, endrin. Slowly degraded are TDE, chlordane, dieldrin, and aldrin. Anaerobic bacteria reduce chlorine present in residues. Diazinon and parathion rapidly degrade by biological hydrolysis. Oxygen-depleted soils often are a hindrance to degradation.

359 Stickel, L[ucille]. F. "Pesticide Residues in Birds and Mammals." In ENVIRONMENTAL POLLUTION BY PESTICIDES, edited by C.A. Edwards, pp. 254-312. London and New York: Plenum Press, 1973.

An able, detailed review covering major aspects of environmental occurrence, physiology, toxicity, and ecosystem kinetics of residues, and the eggshell thinning problem. Excellent reference list.

360 Tucker, Richard K., and Crabtree, D. Glen. HANDBOOK OF TOXICITY OF PESTICIDES TO WILDLIFE. Bureau of Sports Fisheries and Wildlife, Denver Wildlife Research Center, Resource Publication no. 84. Washington, D.C.: Government Printing Office, 1970. 131 p.

Useful summary of toxicological terms, effects, and symptomatology as applied to common laboratory and wildlife species. Most of the book is organized under common names of pesticides.

361 U.S. Department of Health, Education, and Welfare. REPORT OF THE SECRETARY'S COMMISSION ON PESTICIDES AND THEIR RELATIONSHIP TO ENVIRONMENTAL HEALTH. Washington, D.C.: Government Printing Office, 1969. xii, 677 p.

The so-called "Mrak Report" (Emil Mrak, chairman). Still the best coverage of the subject.

362 Van Middelem, C.H. "Fate and Persistence of Organic Pesticides in the Environment." In ORGANIC PESTICIDES IN THE ENVIRONMENT, edited by Aaron A. Rosen and H.J. Kraybill, pp. 228-49. Advances in Chemistry Series, no. 60. Washington, D.C.: American Chemical Society, 1966.

Excellent summary article illustrating title. Charts can be applied to persistent chemicals other than pesticides.

363 Ware, George W., and Roan, Clifford C. "Interaction of Pesticides with Aquatic Microorganisms and Plankton." RESIDUE REVIEWS 33 (1970): 5-45.

Compares studies on soil organisms and aquatic systems. Pesticidal effects on aquatic microorganisms not always predictable. Phenylureas most toxic; cyclodienes most toxic of insecticides. DDT reduces photosynthesis in phytoplankton and is highly toxic to crustaceans. All pesticides are metabolized somewhat. DDT metabolizes to DDE aerobically, to DDD anaerobically.

364 West, Irma. "Biological Effects of Pesticides in the Environment." In ORGANIC PESTICIDES IN THE ENVIRONMENT, edited by Aaron A. Rosen and H.J. Kraybill, pp. 38-53. Advances in Chemistry Series, no. 60. Washington, D.C.: American Chemical Society, 1966.

Good summary article, centering on residues in human beings.

365 Zabik, Matthew J., et al. "Photochemistry of Bioactive Compounds: A Review of Pesticide Photochemistry." ANNUAL REVIEW OF ENTOMOLOGY 21 (1976): 61-79.

Major technical review; includes 191 references. Reports response of pesticides to environmental influences.

Section 5

CHLORINATED HYDROCARBON INSECTICIDES

This class of chemicals is the best known and most widely applied of all insecticides. There is a great variety. Generally they may be characterized by low to moderate toxicity, long persistence in the physical environment and in biological tissues, and by their very low solubility in water but high solubility in fatty materials. They can also be produced at relatively low cost. In use they pose no serious hazard to applicators. Their toxic effects range widely over many groups of organisms although actual toxicity may vary considerably between different kinds of animals.

The chemical qualities that made these compounds so useful in pest control also explain why general broadcast use of them is declining. First, genetic resistance by insects quickly develops after their continued use, aided by the persistence of their residues. Second, both persistence and high fat solubility explain why nontarget organisms accumulated, stored, and passed on in food chains the residues of these chemicals. Third, the general trend toward more selective pest control chemicals and methods, coupled with the prior problems, ensured that chlorinated hydrocarbon insecticides could not remain in general, uncontrolled use. Fourth, when the fears of widespread damage to favored species became supportable by facts, environmentalists acted to initiate new restrictive legislation and regulation. The controversy has been extreme at times. Generally, environmentalists "won the case," but argument continues both in the United States and abroad.

The first part of this section includes titles and annotations which treat chlorinated hydrocarbons generally or which mention more than one chemical compound. Next, several of the better known insecticides are listed by their common names, and articles pertaining specifically to that chemical are included. The longest listing is for DDT, as might be expected. For this compound alone a further division into effects within major groups of organisms is made.

GENERAL AND MULTIPLE REFERENCES

366 Anas, Raymond E., and Wilson, Alfred J. "Organochlorine Pesticides in Nursing Fur Seal Pups." PESTICIDE MONITORING JOURNAL 4

(1970): 114-16.

All twenty-five tissue and ingested milk samples had DDE; nineteen, DDD; twenty-two, DDT; nine, dieldrin. Trace amounts of PCBs were present in all, except brain tissue. Amounts were higher in four-month nursing pups than in four-month fetuses.

367 Barthel, W.F., et al. "Pesticide Residues in Sediments of the Lower Mississippi River and Its Tributaries." PESTICIDES MONITORING JOURNAL 3 (1969): 8-66.

Detailed reporting. Chlorinated hydrocarbon insecticide residues occurred in both agricultural and nonagricultural areas. No indication of buildup.

368 Bennington, Seddon L., et al. "Patterns of Chlorinated Hydrocarbon Contamination in New Zealand Sub-Antarctic and Coastal Marine Birds." ENVIRONMENTAL POLLUTION 8 (1975): 135-47.

New Zealand has been a major user of DDT. Tissues and eggs from coastal birds had high levels of DDT derivatives and of PCBs from industrial sources. In the sub-Antarctic, DDE residues were several times lower and PCB levels lower by one to two orders of magnitude than in species occupying comparable ecological niches at Amchitka on a comparable latitude in the North Pacific. Concentrations of DDE in one egg of a New Zealand falcon were sufficiently high to be associated with eggshell thinning. Excellent survey references.

369 Bevenue, Arthur, et al. "Organochloride Pesticide Residues in Water, Sediment, Algae, and Fish: Hawaii--1970-71." PESTICIDES MONITORING JOURNAL 6 (1972): 56-64.

In low ppt range. Ratio of residues from canal waters in algae, sediment, and fish was 1 : 4000 : 9000 : 32,000. Conclude: little pollution.

370 Bidleman, T.F., and Olney, C.E. "Chlorinated Hydrocarbons in the Sargasso Sea Atmosphere and Surface Water." SCIENCE 183 (1974): 516-18.

In area of Bermuda. Atmospheric chlorinated hydrocarbons gaseous and two orders of magnitude higher than reported earlier in particulate matter. PCB and DDT in surface microlayer (to 30 cm). PCB and DDT in air fall out more rapidly (20x) than previously calculated; that is, enter surface water more rapidly.

371 Brooks, Gerald T. CHLORINATED INSECTICIDES. 2 vols. Cleveland, Ohio: CRC Press, 1974. 249 p; 197 p.

Vol. 1: TECHNOLOGY AND APPLICATIONS; vol. 2 BIOLOGICAL AND ENVIRONMENTAL ASPECTS. Best summary source of all aspects of the origins, analysis, use, residues, and biological problems associated with chlorinated hydrocarbon insecticides. Many references; systematic indexes.

372 Brueggeman, J., et al. "Pesticide and PCB Residue in the Organs of Wild Animals as Indications of Environmental Contamination." ZEITSCHRIFT FUER JAGDWISSENSCHAFT (Hamburg and Berlin) 20 (1974): 70-74. Article in English; journal in German, English, and French.

Eight hundred samples of twenty-six species over a four-year span were checked for residues of chlorinated pesticides and PCBs. Most common were HCB (benzene hexachloride) and lindane (the gamma isomer of BHC). Other pesticides and PCB were five times higher in carnivores than insectivores. Highest in predatory birds and in codfish liver. Much tabulated data.

373 Burns, James E. "Organochlorine Pesticide and Polychlorinated Biphenyl Residues in Biopsied Human Adipose Tissue--Texas, 1967-72." PESTICIDES MONITORING JOURNAL 7 (1974): 122-26.

From 221 surgically derived samples. Total averages: DDT, 23.18 ppm; DDE, 17.37 ppm; dieldrin, 0.35 ppm; gamma BHC, 1.29 ppm. No difference by sex. Mexican Americans had significantly higher DDE, p,p'-DDT, and dieldrin than Anglo-Americans. PCBs uncommon.

374 Butler, Philip A. "Organochlorine Residues in Estuarine Mollusks, 1965-72. National Pesticide Monitoring Program." PESTICIDES MONITORING JOURNAL 6 (1972): 238-362.

Reports development of monitoring program for estuarine mollusks in fifteen coastal states. Over 9,000 samples; tested for fifteen compounds. DDT (max. 5.39 ppm) ubiquitous; dieldrin next common (max. 0.23 ppm).

375 Clark, Donald R., Jr., and McLane, M. Anne Ross. "Chlorinated Hydrocarbon and Mercury Residues in Woodcock in the United States, 1970-71." PESTICIDES MONITORING JOURNAL 8 (1974): 15-22.

Generally low levels of both chemicals, showing declines in relation to earlier surveys. PCBs increased. Mirex levels highest in Louisiana and Mississippi.

376 Clark, Donald R., Jr., et al. "Organochlorine Insecticide Residues in the Free-tailed Bat (_Tadarida brasiliensis_) at Bracken Cave, Texas." JOURNAL OF MAMMALOGY 56 (1975): 429-43.

Residues of DDE in brain of young greater in those on floor compared to those on ceiling; apparently inattention by parent and starvation, not residues, led to fall from ceiling. DDE and DDT highest in yearling pregnant females. Females rapidly excrete residues in milk and nursing young pick up residues quickly. Dieldrin and PCBs present. Data did not indicate adverse effects due to residues. Good bibliography. One would expect residual effects considering food habits and food chain position of the species.

377 Clegg, Donald Edward. "Chlorinated Hydrocarbon Pesticide Residues in Oysters (Crassostrea commercialis) in Moreton Bay, Queensland, Australia--1970-72." PESTICIDE MONITORING JOURNAL 8 (1974): 162-66.

Detailed monitoring results. Variety of residues present, mainly low range.

378 DeLong, Robert L., et al. "Premature Births in California Sea Lions: Association with High Organochlorine Pollutant Residue Levels." SCIENCE 181 (1973): 1168-70.

Premature pupping noted since 1968. Pesticide (total DDT) and PCB residues two to eight times higher in tissues of premature parturient females and pups. Mercury residues in livers in premature females ranged from 38 to 64 ppm.

379 Derr, S. Krogh. "Bioactive Compounds in the Aquatic Environment: Loss of Methoxychlor from Autumn-Shed Leaves into the Aquatic Environment." BULLETIN OF ENVIRONMENTAL CONTAMINATION AND TOXICOLOGY 11 (1974): 500-502.

Oak, maple, and beech retain applied methoxychlor in the order given. Differential suggests detrital effects serve as partitioning reservoir for the chemical.

380 Duke, Thomas W., and Wilson, Alfred J. "Chlorinated Hydrocarbons in Livers of Fishes from the Northeastern Pacific Ocean." PESTICIDE MONITORING JOURNAL 5 (1971): 228-32.

Residue higher from Santa Monica Bay than from pelagic waters. Comparisons also with Baja California. DDE in all, to 2,200 ppm in fat of rockfish from Santa Monica Bay. Dieldrin and PCB recorded twice in traces.

381 Faber, Raymond A., and Hickey, Joseph J. "Eggshell Thinning, Chlorinated Hydrocarbons, and Mercury in Inland Aquatic Bird Eggs, 1969 and 1970." PESTICIDES MONITORING JOURNAL 7 (1973): 27-36.

In upper Great Lakes, nine of thirteen fish-eating bird species have sustained thinning since 1946. Lesser change is evident in herons from Louisiana. On lipid basis, mean PCB and DDE levels exceeded 100 ppm in seven of thirteen species in Great Lakes region; two of seven in Louisiana. Dieldrin values higher in Louisiana. Mercury in eggs exceeded 0.5 ppm in 29 percent of Great Lakes specimens, and 9 percent exceeded 1.0 ppm. Low levels of mercury in Louisiana specimens. PCB has some effect on thinning in mergansers. Detailed tables in text.

382 Faber, Raymond A., et al. "Organochlorines and Mercury in Common Egrets and Great Blue Herons." ENVIRONMENTAL POLLUTION 3 (1972): 111-22.

Reproductive success of a colony of egrets declined during 1967-70. Herons did not parallel this decline, but eggshells for the two species were 15.2 and 10.4 percent thinner respectively than in pre-1947 values. DDE levels comparable to those in other species showing shell thinning. PCB in egrets ranged from 15 ppm in brain, to 93 ppm in liver of adult egrets, and mercury in liver was 2.0 to 9.5 ppm. Dieldrin in brain ranged 5 to 7 ppm--probably the cause of death. Ratios of DDE to PCB indicate that birds acquired residue loads on wintering grounds, not in vicinity of breeding colony.

383 Foehrenbach, Jack, et al. "Chlorinated Hydrocarbon Residues in Shellfish (Pelecypoda) from Estuaries of Long Island, New York." PESTICIDES MONITORING JOURNAL 5 (1971): 243-47.

Reports concentrations of DDT, DDD, DDE, and dieldrin--highest being 0.146 mg/kg, wet weight.

384 Frank, Richard, et al. "Organochlorine Insecticide Residues in Sediment and Fish Tissues, Ontario, Canada." PESTICIDES MONITORING JOURNAL 7 (1974): 165-80.

Authors contrasted areas of heavy, intermediate, and negative use of chemicals. In sediments, all were approximately the same. Residues of DDT and dieldrin in fish depended more on feeding habits, fat, and age than area. Geological origins of sediments a factor.

385 ______ "Residue of Organochlorine Compounds and Mercury in Birds Eggs from the Niagara Peninsula, Ontario." ARCHIVES OF ENVIRONMENTAL CONTAMINATION AND TOXICOLOGY 3 (1975): 205-18.

From twenty species of birds with varied feeding habits; agricultural and urban areas. Aquatic carnivorous birds had highest residues: total DDT, 7.6-22.4 ppm; PCB, 3.5-74.0 ppm; total mercury, 0.64-0.83 ppm. Terrestrial carnivores relatively high: total DDT, 2.5-3.9 ppm; PCB, 0.2-1.0 ppm; mercury, 0.06-0.09 ppm. Herbivores and insectivores much lower values. PCB values generally higher in eggs from urban areas. Similar values, rural and urban, with DDT.

386 Franson, J. Christian, et al. "Chlorinated Hydrocarbon Insecticide Residues in Adipose, Liver, and Brain Samples from Iowa Mink." BUL-

LETIN OF ENVIRONMENTAL CONTAMINATION AND TOXICOLOGY 11 (1974): 379-85.

Study of residue bioconcentration in predatory mammal in wild. Residues of DDT-R to 9.51 ppm, chiefly p,p'-DDE. Dieldrin to 4.90 ppm. Some heptachlor and gamma-BHC. Highest in fatty tissue. Observed levels presumed not to affect reproduction as has been shown (at higher residue levels) in ranch minks.

387 Fredeen, F.J.H., et al. "Residues of Methoxychlor and Other Chlorinated Hydrocarbons in Water, Sand, and Selected Fauna Following Injections of Methoxychlor Black Fly Larvicide into the Saskatchewan River, 1972." PESTICIDES MONITORING JOURNAL 8 (1975): 241-46.

Methoxychlor 0.309 ppm in water picked up many miles downstream; concentrated only briefly in animal tissue. Concentrated in sediments and suspended solids.

388 French, M.C., and Jeffries, D.J. "Degradation and Disappearance of Ortho, Para Isomer of Technical DDT in Living and Dead Avian Tissue." SCIENCE 165 (1969): 914-16.

O,p'-DDT breaks down to p,p'-DDT in living tissue but goes a different route in the anaerobic condition existing after death. Birds not analyzed within a few hours after death will not show o,p'-DDT residues.

389 Gilbertson, Michael. "Seasonal Changes in Organochlorine Compounds and Mercury in Common Terns of Hamilton Harbour, Ontario." BULLETIN OF ENVIRONMENTAL CONTAMINATION AND TOXICOLOGY 12 (1974): 726-32.

Value of organochlorines varies seasonally a great deal; true also of mercury, except for values in eggs. Contaminants came from vicinity of colony; residue levels low upon birds' arrival in spring migration.

390 Gish, Charles D. "Organochlorine Insecticide Residues in Soils and Soil Invertebrates from Agricultural Lands." PESTICIDES MONITORING JOURNAL 3 (1970): 241-52.

Samples collected over eight states included residues of DDE, DDD, DDT, aldrin, dieldrin, endrin, heptachlor epoxide, and chlordane. Organochlorine residues in soils averaged 1.5 ppm, dry weight, and in earthworms 13.5 ppm--a ninefold concentration. Range was from a trace to 19.1 ppm in soils and from a trace to 159.4 ppm in earthworms. Residues in beetle larvae averaged 0.6 ppm; in snails, 3.5 ppm; in slugs, 89.0 ppm. Significant correlations between amounts in soils and in earthworms.

391 Grocki, Darlene R.J., and Johnston, David W. "Chlorinated Hydrocarbon Pesticides in North American Cuckoos." AUK 91 (1974): 186-88.

Authors consider cuckoos as subterminal predators in food chain. Only one of seventeen migratory birds analyzed lacked p,p'-DDT or metabolites. Others had DDT, DDD, and DDE. Dieldrin in one-half but no mirex or PCB in any. Insectivorous birds generally have total DDT load of 5-25 ppm fresh weight.

392 Harris, C.R., and Miles, J.R.W. "Pesticide Residues in the Great Lakes Region of Canada." RESIDUE REVIEWS 57 (1975): 27-79.

Essentially a status report of residues in Ontario, where pesticides are much used and needed. Most uses of DDT, and all use of aldrin, dieldrin, and heptachlor were banned in 1969. Less is known about organophosphorus compounds and carbamates.

393 Harvey, George R., et al. "Chlorinated Hydrocarbons in Open-Ocean Atlantic Organisms." In THE CHANGING CHEMISTRY OF THE OCEANS, edited by David Dyrssen and Daniel Jagner, pp. 177-86. New York: John Wiley & Sons, 1972.

PCBs and DDT found in all organisms sampled from the North Atlantic.

394 Heath, Robert G., and Hill, Sharon A. "Nationwide Organochlorine and Mercury Residues in Wings of Adult Mallards and Black Ducks During 1969-70 Hunting Season." PESTICIDES MONITORING JOURNAL 7 (1974): 153-64.

Order of predominant residue, highest to lowest: DDE, PCB, mercury, DDT, dieldrin, DDD, heptachlor epoxide. Residues highest on coastal flyways. DDE was as high as 41 ppm; slightly higher in males.

395 Henny, Charles J., and Wight, Howard M. "An Endangered Osprey Population: Estimates of Mortality and Production." AUK 86 (1969): 188-98.

Background paper on species responding to pesticide residues by population declines. Acceleration of declines in local populations in mid-Atlantic states since 1940s.

396 Hickey, Joseph J., and Anderson, Daniel W. "Chlorinated Hydrocarbons and Eggshell Changes in Raptorial and Fish-eating Birds." SCIENCE 162 (1968): 271-73.

Catastrophic declines of three raptorial birds in the United States. Declines accompanied by eggshell thinning beginning in 1947, amounting to 19 percent or more. Identical to phenomenon reported in Britain. In 1967 shell thickness in herring gulls directly related to amounts of chlorinated hydrocarbon residues present in eggs.

397 Hickey, Joseph J., et al. "An Exploration of Pesticides in a Lake Michigan Ecosystem." JOURNAL OF APPLIED ECOLOGY 3 (1966): supplement, 141-45.

One of the first sound studies illustrating bioconcentration of residues in a trophic or food chain sequence. Pesticides were originally derived from agricultural lands adjacent to Lake Michigan.

398 Holden, A.V. "International Cooperative Study of Organochlorine Pesticide Residues in Terrestrial and Aquatic Wildlife, 1967/1968." PESTICIDES MONITORING JOURNAL 4 (1970): 117-35.

Collaborative studies from seventeen laboratories in eleven countries reported. Range of variation in values high in natural populations. Large samples required to ensure reliability of conclusions.

399 Holden, A.V., and Marsden, K. "Organochlorine Pesticides in Seals and Porpoises." NATURE 216 (1967): 1274-76.

In Scotland and Canada, far from sites of application, seals and porpoises accumulate high concentrations of pesticide residue in blubber. Residue spreads through long food chain, is dispersed to distant places, and ends with large marine mammals. Residues include dieldrin, p,p'-DDE, p,p'-DDD, and p,p'-DDT. DDE and DDT in porpoise are the highest, with concentrations to 25.7 ppm.

400 Johnson, Donald W., and Lew, Sam. "Chlorinated Hydrocarbon Pesticides in Representative Fishes of Southern Arizona." PESTICIDE MONITORING JOURNAL 4 (1970): 57-61.

Most residues are in ppb range. DDT and metabolites to 187.5 ppm. Toxaphene as great as 172.9 ppm. DDE commonest. Source is wastewaters from croplands entering Colorado River.

401 Kapoor, Indor P., et al. "Comparative Metabolism of Methoxychlor, Methiochlor, and DDT in Mouse, Insects, and in a Model Ecosystem." JOURNAL OF AGRICULTURAL AND FOOD CHEMISTRY 18 (1970): 1145-52.

Radiolabeled compounds given orally to mice resulted in elimination within twenty-four hours of 98.3 percent of methoxychlor, 47.11 percent of methiochlor, and 1.02 percent of DDT. In a model ecosystem, methoxychlor was found in fish at the top of the food chain at level of 1,500 times that of water, compared to 90,000 times for DDT. Methiochlor found only in trace amounts in fish. DDT, DDD, DDE are clearly not rapidly eliminated by excretory systems.

402 Koeman, J.H., and Van Genderen, H. "Some Preliminary Notes on Residues of Chlorinated Hydrocarbon Insecticides in Birds and Mammals

in the Netherlands." JOURNAL OF APPLIED ECOLOGY 3 (1966): supplement, 99-106.

Birds of prey and fish-eating birds have accumulated large amounts of these residues. High levels may have contributed to observed population declines in spoonbills and terns. Dieldrin and DDT found in subcutaneous fat of harbor seals. Total DDT ranges from 9.6 to 27.4 ppm in blubber; dieldrin, 0.3 to 2.3 ppm.

403 Kreitzer, J.F. "Residues of Organochlorine Pesticides, Mercury and PCB's in Mourning Doves from Eastern United States, 1970-71." PESTICIDES MONITORING JOURNAL 7 (1974): 195-99.

Residues of DDT-R averaged 5.0 ppm, lipid weight; 0.068 ppm, wet weight. PCB residues averaged 9.75 ppm, lipid weight; 0.121 ppm, wet weight. Other residues in traces.

404 Lamont, Thair, and Reichel, William. "Organochlorine Pesticide Residues in Whooping Cranes and Everglade Kites." AUK 87 (1970): 158-59.

Both species rare and protected. Low levels of residues of DDT series and dieldrin in tissues and in one egg of a kite. Also low levels (maximum 0.06 ppm) of residue in apple snails (*Pomacea paludosa*), the primary food of Everglade kites. Significance of residues not known.

405 Law, LeRoy M., and Goerlitz, D.F. "Selected Chlorinated Hydrocarbons in Bottom Material from Streams Tributary to San Francisco Bay." PESTICIDES MONITORING JOURNAL 8 (1974): 33-36.

Good study. Ubiquitous in twenty-six streams were chlordane, DDD, DDE, DDT, and PCB. One stream had polychlorinated naphthalene residues.

406 Lincer, Jeffrey L., and Sherburne, James A. "Organochlorines in Kestrel Prey: A North-South Dichotomy." JOURNAL OF WILDLIFE MANAGEMENT 38 (1974): 427-34.

Microtine rodents were most common prey items as indicated by regurgitated pellets. These rodents had low DDE values (0.01 ppm wet weight) and even with residues in other prey species cannot explain high values in eggs. Authors postulated that kestrels picked up residues on southern wintering grounds.

407 Lincer, Jeffrey L., et al. "Organochlorine Residues in Alaskan Peregrine Falcons (*Falco peregrinus* Tunstall), Rough-legged Hawks (*Buteo lagopus* Pontoppidan) and Their Prey." CANADIAN FIELD-NATURALIST 84 (1970): 225-64.

Resident Arctic prey have less than 1.0 ppm total residue in diet; migratory bird prey have ten to twenty times as much residue as resident small mammals commonly preyed upon by rough-legged hawks.

408 Lu, Po-Yung, et al. "Evaluation of Environmental Distribution and Fate of Hexachlorocyclopentadiene, Chlordene, Heptachlor, and Heptachlor Epoxide in a Laboratory Model Ecosystem." JOURNAL OF AGRICULTURAL AND FOOD CHEMISTRY 23 (1975): 967-73.

The problem is the rate of loss of residue by hydroxylation in chlordene, heptachlor, and heptachlor epoxide. The last named is a very stable toxic compound; i.e., it degrades slowly.

409 McLane, M. Anne Ross, et al. "Organochlorine Pesticide Residues in Woodcock, Soils, and Earthworms in Louisiana, 1965." PESTICIDES MONITORING JOURNAL 5 (1971): 248-50.

Samples taken three years after termination of aerial heptachlor treatments in this region. Epoxide residues in woodcock averaged 0.42 ppm (dry weight), much lower than 1961-62 values. Residues of DDE averaged 3.62 ppm, higher than 1961-62 values. Earthworms and soils contained traces of several organochlorine pesticides.

410 Menzel, David W., et al. "Marine Phytoplankton Vary in Their Response to Chlorinated Hydrocarbons." SCIENCE 167 (1970): 1724-26.

Four marine phytoplankton species exposed to DDT, dieldrin, and endrin varied in responses from insensitivity in *Dunaliella*, to toxicity in *Cyclotella*, at 0.1-1.0 ppb. Other forms intermediate in response.

411 Metcalf, Robert L., et al. "Model Ecosystem Studies of the Environmental Fate of Six Organochlorine Pesticides." ENVIRONMENTAL HEALTH PERSPECTIVES, Experimental Issue no. 4 (1973): 35-44.

Report of model ecosystems in laboratories employing, in this study, radio-marked pesticides (dieldrin, endrin, mirex, hexachlorobenzene, lindane, and DDT). Biomagnification in these pesticides dramatically shown. Residues accumulated from 200 to 84,000 times in fishes and snails after thirty-three days. These rates contrast sharply with rates in organophosphate compounds. The value of the model ecosystem as a registration technique is illustrated.

412 Modin, John C. "Chlorinated Hydrocarbon Pesticides in California Bays and Estuaries." PESTICIDES MONITORING JOURNAL 3 (1969): 1-7.

DDT, DDD, DDE, dieldrin, and endrin residues in oysters, mussels, and clams were from 10 to 3,600 ppb, wet weight. Offshore, concentrations of DDT, DDD, and DDE were 2,739 ppb in crab; 668 ppb in ova from King crab; 591 ppb in ova of fishes, including halibut.

413 Moriarty, F[rank]., ed. ORGANOCHLORINE INSECTICIDES: PERSIS-

TENT ORGANIC POLLUTANTS. New York: Academic Press, 1975. xii, 302 p.

Unusual in that the book attempts to identify general principles to help predict and control events. Six sections, individually authored. Contents: A.V. Holden, "Monitoring Persistent Organic Pollutants"; F. Moriarty, "Exposures and Residues"; C.H. Walker, "Variations in the Intake and Elimination of Pollutants"; D.J. Jeffries, "The Role of the Thyroid in the Production of Sublethal Effects by Organochlorine Insecticides and Polychlorinated Biphenyls"; T.P. Dempster, "Effects of Organochlorine Insecticides on Animal Populations"; P.A. Victor and W.M. Mansell, "Persistent Pesticides: An Economic and Legal Analysis." British authors.

414 Nash, Ralph G., and Woolson, Edwin A. "Persistence of Chlorinated Hydrocarbons in Soils." SCIENCE 157 (1967): 924-27.

Percentages of substances remaining in sandy loam soil after many years were technical (commercial grade) aldrin, 40; chlordane, 40; endrin, 41; heptachlor, 16; dilan, 23; isodrin, 15; BHC, 10; and toxaphene, 45. Percentage of DDT in three soil types after seventeen years was 39. Values approach an upper limit to insecticide persistence in soil.

415 Newton, I., and Bogan, J. "Organochlorine Residues, Eggshell Thinning and Hatching Success in British Sparrowhawks." NATURE 249 (1974): 582-83.

Main residues are DDE, PCB, and dieldrin. Widespread shell thinning. Failures of clutches include no nest building, not laying, egg breakage, desertions, and embryo deaths. Significantly higher organochlorine levels in birds showing these types of failures than those with normal success.

416 Parslow, J.L.F., and Jeffries, D.J. "Relationship between Organochlorine Residues in Livers and Whole Bodies of Guillemots." ENVIRONMENTAL POLLUTION 5 (1973): 87-101.

Massive die-off of birds in Irish Sea in 1969. Birds found dead showed significant relationship between liver and whole body residues of DDE and PCB, but not of dieldrin. Each bird had about 4,660 micrograms of PCBs, about twice as high as healthy birds. PCB and natural stresses probably implicated in this avian diaster in which tens of thousands of birds died.

417 Peakall, David B., et al. "Organochlorine Residues in Alaskan Peregrines." PESTICIDES MONITORING JOURNAL 8 (1975): 255-60.

Residues in eggs maintained essentially constant level from 1969 to 1973 despite decreased use of pesticides in the United States and Canada. If total residues are over 20 ppm, population declines can be expected. This decline is now evident in Colville River area, Alaska, where last falcon will fledge in 1975 and species is predicted to be gone by 1980. Decline is not predictable at lower egg residue levels. At 5-7 ppm, as in Aleutian Islands, reproduction remained normal and population level was maintained.

418 Presstt, Ian. "Organochlorine Pollution of Rivers and the Heron (Ardea cinerea L.)." In PAPERS AND PROCEEDINGS 11TH TECHNICAL MEETING, INTERNATIONAL UNION FOR CONSERVATION OF NATURE AND NATURAL RESOURCES, pp. 95-102. Morges, Switzerland: 1970.

Paper from the conference held at New Delhi, 25-28 November 1969, reports on values of DDE, DDT, dieldrin, and PCB in liver and eggs of herons in Britain. (Heron numbers in Britain are closely monitored.) Through accidental contamination of rivers in eastern England, residues of dieldrin and p,p'-DDE have built up in food chains to a point where all fish are contaminated. In trout, a nearly lethal level exists; in herons, levels are lethal to some, and have probably interfered with breeding capacity. The heron is a good pollution indicator in river systems.

419 Prouty, Richard M., et al. "DDD Poisoning in a Loon and the Identification of the Hydroxylated Form of DDD." BULLETIN OF ENVIRONMENTAL CONTAMINATION AND TOXICOLOGY 14 (1975): 385-88.

DDD presumed responsible for death because level in brain was 200 ppm. Specimen found 1.5 years after moratorium on DDT and DDD use.

420 Ratcliffe, D.A. "Decrease in Eggshell Weight in Certain Birds of Prey." NATURE 215 (1967): 208-10.

Key paper historically. Increase in incidence of broken eggs in British birds of prey (peregrine, sparrowhawk, and golden eagle) since 1950. Author links behavioral changes, eggshell thinning, decrease in egg weight, increase in organochlorine residues, and altered circulating calcium levels. This article should be read in relation to later papers on hepatic enzyme induction; steroid hormones.

421 Reichel, William L., et al. "Pesticide Residues in Eagles." PESTICIDES MONITORING JOURNAL 3 (1969): 142-44.

DDE, DDD, and dieldrin found in all samples. Bald eagles have higher levels than golden eagles. A few had residues of heptachlor expoxide and endrin.

422 Reidinger, Russell F., Jr., and Crabtree, D. Glen. "Organochlorine Residues in Golden Eagles, United States--March 1964-July 1971."

PESTICIDES MONITORING JOURNAL 8 (1974): 37-43.

All 169 birds had residues. DDE and PCBs commonest. Great variability but no regional patterns. Food intake must be source. No evidence of residues causing death.

423 Risebrough, Robert W. "Chlorinated Hydrocarbons in Marine Ecosystems." In CHEMICAL FALLOUT, pp. 5-23. Springfield, Ill.: Charles C. Thomas, Publisher, 1969.

Residue values of DDT compounds and PCBs in marine fishes and birds. Commentary on residue dispersal mechanisms in air and water.

424 Risebrough, Robert W., et al. "DDT Residues in Pacific Sea Birds: A Persistent Insecticide in Marine Food Chains." NATURE 216 (1967): 589-91.

The earliest paper to show the extent of chlorinated hydrocarbon residue dispersal through marine ecosystems. Aerial dispersion followed by biological concentration discussed.

425 ______ "Organochlorine Pollutants in Peregrines and Merlins Migrating through Wisconsin." CANADIAN FIELD-NATURALIST 84 (1970): 247-53.

DDD, DDE, PCB, dieldrin levels in both peregrines and merlins are very high. Chemicals similar to PCB require watching.

426 ______. "Pesticides: Transatlantic Movements in the Northeast Trades." SCIENCE 159 (1968): 1233-36.

Concentrations of chlorinated hydrocarbons in airborne dust carried by trade winds from European-African land mass to Barbados range are from less than 1.0 to 164 ppb. Pesticides contributed to tropical waters by trade winds comparable to those carried to the sea by major river systems.

427 ______. "Reproductive Failure of the Brown Pelican on Anacapa Island in 1969." AMERICAN BIRDS 25 (1971): 8-9.

Reproductive failure of the brown pelican on Anacapa Island (southern California) in 1969 is presumed due to high levels of chlorinated hydrocarbon residues. Of 1,272 nests only two produced a single young apiece. Eggs were produced by most birds but there was high egg breakage and mortality of young.

428 Robel, R.J., et al. "Effects of Insecticides on Populations of Rodents in Kansas--1965-69." PESTICIDES MONITORING JOURNAL 6 (1972): 115-21.

Long-term study. Heavily treated agricultural area was compared to untreated area. No important difference in population dynamics of the deer mouse, *Peromyscus maniculatus*.

429 Roberts, D. "Sub-lethal Effects of Chlorinated Hydrocarbons in Bivalves." MARINE POLLUTION BULLETIN 6 (1975): 20-24.

Chlorinated hydrocarbon pesticides are now present in low concentration in most waters. Generally do not cause direct mortality so far as known; may have sublethal effects, reducing health and viability of marine organisms. Roberts studied effects of endosulfan, DDT, and dieldrin on respiration, heartbeat, and burrowing activity.

430 Robinson, J. "Dynamics of Organochlorine Insecticides in Vertebrates and Ecosystems." NATURE 215 (1967): 33-35.

Mathematical expressions based on set theory, which indicate that research based on measurement of pesticides at different periods of time can lead to misleading conclusions. Mainly based on tissue distributions of residues.

431 Schneeweis, James C., et al. "Organochlorine Pesticide Residue Levels in North American Timber Wolves--1969-71." PESTICIDES MONITORING JOURNAL 8 (1974): 142-43.

Fifty-one wolves analyzed for nine chemicals. Conclude: very low contamination in wolves and their environment in Ontario and northern Minnesota.

432 Shaw, Stanton B. "Chlorinated Hydrocarbon Pesticides in California Sea Otters and Harbor Seals." CALIFORNIA FISH AND GAME 57 (1971): 290-94.

Highest levels of DDT, DDD, DDE, and total DDT were found in fat of both species. In sea otters total residues ranged from 0.41 to 36 ppm, wet weight; in seals 18-158 ppm, wet weight.

433 Snyder, N., et al. "Organochlorines, Heavy Metals, and the Biology of North American Accipiters." BIOSCIENCE 23 (1973): 300-305.

DDE is the prime residue pollutant of accipiters in the southwestern United States. Heavy metals and PCB are present at low levels but effects are unknown. Levels of cadmium from unsuccessful nests are higher than from successful.

434 Vermeer, Kees, and Reynolds, Lincoln C. "Organochlorine Residues in Aquatic Birds in the Canadian Prairie Provinces." CANADIAN FIELD-NATURALIST 84 (1970): 117-30.

Twenty-one species in Alberta, Saskatchewan, and Manitoba. DDE and dieldrin levels higher in eggs of gulls and fish-eating birds than in ducks and geese. Feeding habits (difference in trophic position) dictate tissue levels. DDE, dieldrin, and PCB levels in gull tissues predictable when egg values known. Shell thickness significantly and inversely correlated with DDE concentrations in blue heron eggs, but no correlation with PCB.

435 Walsh, F., and Mitchell, R. "Inhibition of Inter-microbial Predation by Chlorinated Hydrocarbons." NATURE 249 (1974): 673-74.

Experiments with Escherichia coli and their marine predators in presence of 2,4-D and o,o, dichlorophenyl. Lower concentrations significantly reduced kill rate of E. coli. Residues may decrease level of predator-prey interactions responsible for self-purification in natural ecosystems in vicinity of outfalls.

436 Wiemeyer, Stanley N., et al. "Residues of Organochlorine Pesticides, Polychlorinated Biphenyls and Mercury in Bald Eagle Eggs and Changes in Shell Thickness--1969 and 1970." PESTICIDES MONITORING JOURNAL 6 (1972): 50-55.

All eggs contained residues of DDE, dieldrin, PCBs, and mercury. Lowest values in eggs from Alaska. Most showed eggshell thinning. Some eggs had DDE levels comparable to experimental situation in which thinning occurred. High dieldrin residues could be hampering reproductive success.

437 Wolman, Allen A., and Wilson, Alfred J. "Occurrence of Pesticides in Whales." PESTICIDES MONITORING JOURNAL 4 (1970): 8-10.

Six of twenty-three gray whales and all six sperm whales sampled near San Francisco had organochlorine residues. DDT and metabolites were up to 0.36 ppm in gray whale blubber and to 6.0 ppm in sperm whale. Highest concentration was dieldrin, 0.075 ppm in gray and 0.019 ppm in sperm whales.

438 Wurster, C[harles]. F., Jr. "Chlorinated Hydrocarbon Insecticides and Avian Reproduction: How Are They Related?" In CHEMICAL FALLOUT, pp. 368-407. Springfield, Ill.: Charles C. Thomas, Publisher, 1969.

Residues of hydrocarbons, particularly DDT, interfere with calcium metabolism and hormonal mechanisms. They are particularly damaging in carnivorous birds.

CYCLODIENE INSECTICIDES

Most uses of the cyclodiene insecticides are now banned by federal legislation. Nonetheless, they had been in common use from the late 1940s until their prohibition. As a class of chemicals, they were moderately toxic, most persistent, and relatively inexpensive. Their immediate toxicity was sufficiently high to cause a great deal of environmental damage, particularly documented by fish and wildlife losses. Insect resistance rapidly followed their broadcast use. Two features were particularly complicating: the conversion of aldrin to the more toxic dieldrin under field conditions; and the conversion under sunlight to the photoisomers which were more toxic than the parent chemical. Moreover, dieldrin has been shown to be carcinogenic under experimental conditions. Much controversy has accompanied the use of these chemicals.

439 Batterton, J.C., et al. "Growth Response of Blue-Green Algae to Aldrin, Dieldrin, Endrin and Their Metabolites." BULLETIN OF ENVIRONMENTAL CONTAMINATION AND TOXICOLOGY 6 (1971): 589-94.

Metabolic products of parent chemicals can be at least as inhibitory to algal growth as parent compounds.

440 Baxter, William L., et al. "Dieldrin Effects in Two Generations of Penned Hen Pheasants." JOURNAL OF WILDLIFE MANAGEMENT 33 (1969): 96-102.

Pheasants were fed up to 6 mg dieldrin in capsules in thirteen weeks. No mortality in first generation but there was mortality in all groups in second generation. Food consumption and egg production lower in both generations. Fertility and hatchability much lower in second generation birds that received dieldrin through the egg. Dieldrin levels in brain tissues of birds that died ranged to 26.7 ppm. Residues in wild birds lower than residues in penned birds.

441 Bridges, William R. "Disappearance of Endrin from Fish and Other Materials of a Pond Environment." TRANSACTIONS OF THE AMERICAN FISHERIES SOCIETY 90 (1961): 332-34.

Bridges reports on the toxicity and the survival time of endrin in mud, water, fish, and aquatic vegetation. Endrin takes two to three months to "disappear"; dilution is one factor.

442 Brooks, Gerald T. "The Metabolism of Diene-Organochlorine (Cyclodiene) Insecticides." RESIDUE REVIEWS 27 (1969): 81-138.

Excellent referencing and review of a class of compounds not now in use in the United States. Historically important and still widely used outside of United States. Good section on bird and mammal responses.

443 Bunch, Thomas D., and Low, Jessop B. "Effects of Dieldrin on Chromosomes of Semi-domestic Mallard Ducks." JOURNAL OF WILDLIFE MANAGEMENT 37 (1973): 51-57.

Ducks fed up to 30 ppm dieldrin in sixty days. Only at highest levels were chromosomal aberrations shown. The authors conclude that chromosomal effects are not likely at levels normally found in waterfowl.

444 Call, D.J., and Harrell, B.E. "Effects of Dieldrin and PCB's upon the Production and Morphology of Japanese Quail Eggs." BULLETIN OF ENVIRONMENTAL CONTAMINATION AND TOXICOLOGY 11 (1974): 70-77.

Significant eggshell thinning with all dieldrin and PCB treatments. Eggs shorter and narrower on Arochlor. Egg production lessened. Severe weight loss and mortality at higher PCB levels. Good review bibliography.

445 Chadwick, George G., and Brocksen, Robert W. "Accumulation of Dieldrin by Fish and Selected Fish-Food Organisms." JOURNAL OF WILDLIFE MANAGEMENT 33 (1969): 693-700.

Fish in contaminated water (to 8.6 ppb) took up dieldrin from both water and food. Uptake is slower by naturally contaminated foods than experimentally contaminated.

446 Dahlgren, Robert B., and Linder, Raymond L. "Effects of Dieldrin in Penned Pheasants through the Third Generation." JOURNAL OF WILDLIFE MANAGEMENT 38 (1974): 320-30.

Offspring of parents having received dieldrin are more easily caught by hand; more often than the controls, these pheasants elected deep side of "visual cliff." These effects are more pronounced when both parents and grandparents had received dieldrin, implying genetic mechanism and continuing effects of poison.

447 _____. "Eggshell Thickness in Pheasants Given Dieldrin." JOURNAL OF WILDLIFE MANAGEMENT 34 (1970): 226-28.

Up to 10 mg/week dieldrin in diet produced no changes.

448 Dahlgren, Robert B., et al. "Dieldrin Effects on Susceptibility of Penned Pheasants to Hand Capture." JOURNAL OF WILDLIFE MANAGEMENT 34 (1970): 957-59.

Young birds that had received dieldrin in diet were more easily caught, an effect exhibited more if both parents had received dieldrin, less if only hens or cocks (in that order) had received the chemical.

449 Davison, Kenneth L., and Sell, Jerry L. "Dieldrin and DDT Effects on Reproduction and Some Hepatic Mixed-Function Oxidases in the Mallard Duck." ARCHIVES OF ENVIRONMENTAL CONTAMINATION AND TOXICOLOGY 2 (1974): 302-14.

Long-term (343 days) feeding study including two laying seasons showed 200 ppm DDT to be lethal. Tremors occurred in young of second year whose parents received 10 ppm dieldrin. Liver effects not apparent. Residues variable in eggs but higher in second season and were dose-dependent.

450 _____. "Dieldrin and p,p'-DDT Effects on Some Microsomal Enzymes of Liver of Chickens and Mallard Ducks." JOURNAL OF AGRICULTURAL AND FOOD CHEMISTRY 20 (1972): 1198-1205.

Both chemicals in low amounts induced some microsomal enzymes associated with metabolism of estradiol. Enzyme induction in these experiments should be related to the controversy on eggshell thinning as consequence of the altered calcium physiology caused by pesticide residues.

451 Edmundson, Walter F., et al. "Dieldrin Storage Levels in Necropsy Adipose Tissue from a South Florida Population." PESTICIDES MONITORING JOURNAL 2 (1968): 86-89.

Mean value of dieldrin levels was 0.23 ppm. Dieldrin did not stratify by age, sex, or race as did DDT and DDE.

452 Epstein, Samuel S. "The Carcinogenicity of Dieldrin." SCIENCE OF THE TOTAL ENVIRONMENT (Amsterdam) Part I - 4 (1975): 1-52; Part II - 4 (1975): 205-17.

Expert witness's testimony on hearings to ban aldrin and dieldrin confirms that dieldrin induces cancer in experimental animals and probably does in man. Several factors make it difficult to estimate carcinogenicity of widespread synthetic chemicals in man: no ability to experiment; the multifactorial exposure to man; the possibly long latent period before expression; the necessarily short period of time that research on experimental animals takes relative to possible effects in man.

453 Fikes, Martha H., and Tubb, Richard A. "Dieldrin Uptake in the Three-ridge Naiad." JOURNAL OF WILDLIFE MANAGEMENT 36 (1972): 802-9.

A freshwater mollusc known to be a bioconcentrator and pollution indicator was exposed to a continuous background level of less than 20 ppb dieldrin. Release rate was one-sixth that of uptake. Gills showed concentration to 700 ppb; therefore gill structures can be used for monitoring dieldrin in water.

454 Flickinger, Edward L., and King, Kirke A. "Some Effects of Aldrin-treated Rice on Gulf Coast Wildlife." JOURNAL OF WILDLIFE MANAGEMENT 36 (1972): 706-27.

A detailed and important paper. Many tabulations. Particular importance is attached to the precipitous decline of the fulvous tree duck (Dendrocygna bicolor) which, already rare, has suffered from aldrin-dieldrin poisoning. Invertebrates from rice fields where ducks feed showed average values of 9.5 ppm aldrin-dieldrin. Ducks died with brain tissue values below 6.8 ppm dieldrin. Aldrin added to seed rice in this manner is dangerous to wildlife.

455 Fowler, J.F., et al. "Effect of Dieldrin on Egg Hatchability, Chick Survival and Eggshell Thickness in Purple and Common Gallinules." BULLETIN OF ENVIRONMENTAL CONTAMINATION AND TOXICOLOGY 6 (1971): 495-501.

Gallinule eggs taken from areas in Louisiana in which seed rice was treated with aldrin were incubated. Seed had 2500 ppm aldrin. Aldrin converts to the more toxic dieldrin in the field. Dieldrin was found in all analyses. Hatchability was unaffected compared to controls and there was no eggshell thinning in spite of high dieldrin residues in eggs.

456 Gannon, Norman, and Decker, G.L. "The Conversion of Aldrin to Dieldrin on Plants." JOURNAL OF ECONOMIC ENTOMOLOGY 51 (1958): 8-11.

One of the earliest studies to indicate that aldrin converts to the more toxic dieldrin in the field.

457 Hall, J.E., et al. "Effects of Aldrin on Young Pen-reared Pheasants." JOURNAL OF WILDLIFE MANAGEMENT 35 (1971): 429-34.

Weekly administration of aldrin of 1.0 mg or 1.5 mg depressed growth rate. Dieldrin (converted from aldrin) increased in brain tissue in proportion to dose, but aldrin did not. At termination of experiments dieldrin levels in surviving birds were nearly the same as control levels. One-half of birds, fed single dose of 1.5 mg, died within two days.

458 Khan, H.M., and Khan, M.A.Q. "Biological Magnification of Photodieldrin by Food Chain Organisms." ARCHIVES OF ENVIRONMENTAL CONTAMINATION AND TOXICOLOGY 2 (1974): 289-301.

There are marked species differences in absorption and retention of the sunlight conversion product in fresh water in invertebrates and fishes. Crayfish and clam low; mosquito larvae, amphipod, cladoceran, and daphnia high. Levels five to six times higher in guppy and goldfish than in minnow and bluegill. Maximum biological absorption ratios (water to animal): bluegill, 133; minnow, 150; goldfish, 609; guppy, 620; cladoceran, 1,000; amphipod, 1,200; and daphnia, 63,000.

459 Khan, M.A.Q. "In Vivo and in Vitro Epoxidation of Aldrin by Aquatic Food Chain Organisms." BULLETIN OF ENVIRONMENTAL CONTAMINATION AND TOXICOLOGY 8 (1972): 219-28.

Aldrin converts to dieldrin in living organisms; low amounts in algae, high in protozoans. Many other groups considered; largest capacity to epoxidize aldrin found in mosquito larvae and dragonfly nymphs.

460 Khan, M.A.Q., et al. "Toxicity-Metabolism Relationship of the Photoisomers of Certain Chlorinated Cyclodiene Insecticide Chemicals." ARCHIVES OF ENVIRONMENTAL CONTAMINATION AND TOXICOLOGY 1 (1973): 159-69.

Photoisomers of aldrin, dieldrin, and heptachlor are more toxic to freshwater invertebrates, vertebrates, and house flies than parent compounds are. They are also more lipophilic. Therefore it is important to have more research on food chain organisms.

461 Korschgen, Leroy J. "Disappearance and Persistence of Aldrin after Five Annual Applications." JOURNAL OF WILDLIFE MANAGEMENT 35 (1971): 494-500.

Ninety-five percent of pesticide had disappeared before study began. However, after six years, residues had not declined in earthworms, which contained four to fifteen times as much aldrin plus dieldrin as found in soils. (0.11 ppm in soils; 0.61 in earthworms are averages.)

462 ______ "Soil-Food-Chain-Pesticide Wildlife Relationships in Aldrin-treated Fields." JOURNAL OF WILDLIFE MANAGEMENT 34 (1970): 186-99.

Korschgen studied relationships in fields treated many years with 1.0 lb/acre aldrin. Most residues found as dieldrin (to which aldrin converts). Residue averages: soil, 0.31 ppm; earthworms, 1.39 ppm; crickets, 0.23 ppm; ground beetles, 1.10 ppm and in other species of beetles, 9.67 ppm; white-footed mice, 0.98 ppm; toads, 3.53 ppm; garter snakes, 12.35 ppm; crop seeds, less than 0.02 ppm. An exceptional case occurred in a predaceous beetle, *Poecilus*, where concentration was 37.8 ppm. Biological concentration after long-term treatment clearly shown.

463 Krieger, R.I., and Lee, P.W. "Inhibition of in Vivo and in Vitro Epoxidation of Aldrin, and Potentiation of Toxicity of Various Insecticide Chemicals by Diquat in Two Species of Fish." ARCHIVES OF ENVIRONMENTAL CONTAMINATION AND TOXICOLOGY 1 (1973): 112-32.

Diquat, a patent contact herbicide for aquatic weeds, inhibits aldrin epoxidation. Dieldrin conversion is much less when Diquat is present. Detailed paper; high, impractical concentrations of Diquat required to inhibit epoxidation or to potentiate toxicity.

464 Lehner, P[hilip]. N. "Failure of Dieldrin to Affect Color Vision in Juvenile Mallard Ducks." ARCHIVES OF ENVIRONMENTAL CONTAMINATION AND TOXICOLOGY 1 (1973): 303-11.

Ducks on 0, 4, or 10 ppm in diet were trained to give "key-peck" response at different wavelengths of light. Apparently dieldrin has no effect on ability to discriminate wavelengths.

465 Lehner, Philip N., and Egbert, Allen. "Dieldrin and Eggshell Thickness in Ducks." NATURE 224 (1969): 1218-19.

Long-term ingestion of dieldrin (up to 10 ppm) in diet resulted in statistically significant reduction in eggshell thickness in penned mallards. Sublethal doses cause about 4 percent reduction.

466 Lichtenstein, E. P[aul]., et al. "Effects of a Cover Crop versus Soil Cultivation on the Fate and Vertical Distribution of Insecticide Residues in Soil 7 to 11 Years after Soil Treatment." PESTICIDES MONITORING JOURNAL 5 (1971): 218-22.

Soil cultivation resulted in 76-82 percent reduction in residues compared to effect of noncultivated crop (alfalfa). Authors discuss vertical distribution of aldrin, dieldrin, and heptachlor.

467 Linder, Raymond L., et al. "Residues in the Brain of Adult Pheasants Given Dieldrin." JOURNAL OF WILDLIFE MANAGEMENT 34 (1970): 954-56.

Apparently residue level of 3-4 ppm or greater in brain tissue may be taken as values imputing death caused by dieldrin.

468 Lindstrom, F.T., et al. "Distribution of Heod (Dieldrin) in Mammals. I. Preliminary Model." ARCHIVES OF ENVIRONMENTAL CONTAMINATION AND TOXICOLOGY 2 (1974): 9-42.

Mathematical model based on lipid-phase transport, and liver enzyme destruction of dieldrin. Complicated paper with long bibliography.

469 Ludke, J. Larry. "Interaction of Dieldrin and DDE Residues in Japanese Quail (*Coturnix coturnix japonica*)." BULLETIN OF ENVIRONMENTAL CONTAMINATION AND TOXICOLOGY 11 (1974): 297-302.

DDE levels rose faster in tissues and reached a higher plateau when dieldrin (at 1 ppm in diet) was also administered. This article provides an example of synergism, an important idea, but one that is difficult to assess when we find many different kinds of residue pollutants in field-taken specimens.

470 Matsumura, F[umio]., and Boush, G. M[allory]. "Dieldrin: Degradation by Soil Microorganisms." SCIENCE 156 (1967): 959-61.

Dieldrin is the most stable and hazardous of the chlorinated hydrocarbons in the environment, but there is microbial action to facilitate degradation.

471 Moore, N[orman]. W., and Tatton, J. O'G. "Organochlorine Insecticide Residues in the Eggs of Sea Birds." NATURE 207 (1965): 42-43.

Dieldrin and p,p'-DDE found in all eggs of thirteen species. Same for fish and molluscs in region of sea bird colonies in British Isles. Residue levels should not affect hatching success, however.

472 Moriarty, F[rank]. "Residues in Animals during Chronic Exposure to Dieldrin." In GLOBAL ASPECTS OF CHEMISTRY, TOXICOLOGY AND TECHNOLOGY AS APPLIED TO THE ENVIRONMENT, edited by F. Coulston and F. Korte, vol. 3, pp. 104-12. New York: Academic Press, 1974.

After chronic feeding, chlorinated hydrocarbon residues in the tissues of animals normally level off at a predictable plateau. In this study, after eighteen months exposure residue levels suddenly ascended to a second, higher plateau. Reasons for this are not known although alternative hypotheses are discussed.

473 Muller, H.D., and Lockman, D.C. "Fecundity and Progeny Growth Following Subacute Insecticide Ingestion by the Mallard." POULTRY SCIENCE 51 (1972): 239-41.

Mallard ducks fed ad libitum either 4.0 ppm dieldrin or 10.0 ppm parathion in diet. No effects on egg production and fertility. Hatchability depressed 50 percent in dieldrin-fed. Eggshell thickness significantly depressed in parathion-fed; slightly depressed in dieldrin-fed. No effects on progeny.

474 Murphy, Dean A., and Korschgen, Leroy J. "Reproduction, Growth and Tissue Residues of Deer Fed Dieldrin." JOURNAL OF WILDLIFE MANAGEMENT 34 (1970): 887-903.

Over a three-year period, dieldrin (to 25 ppm) in diet produced no toxic effects in deer, including progeny. But fawns were smaller and their growth slower. Fawns had more residues in brain tissues than older animals when their mothers' milk contained up to 12 ppm of dieldrin. Dieldrin in the diet of fawns was fatal at 100 and 200 ppm levels within twenty-seven days.

475 Neill, D.D., et al. "Pesticide Effects on the Gray Partridge." BULLETIN OF ENVIRONMENTAL CONTAMINATION AND TOXICOLOGY 6 (1971): 546-51.

Partridge (*Perdix perdix*) were fed ad libitum 3.0 ppm dieldrin and 8.0 ppm parathion. No parathion in eggs but high embryo death. No effects on growth rate with either chemical. Mortality low in chicks.

476 Ottolenghi, Anne D., et al. "Teratogenic Effects of Aldrin, Dieldrin,

and Endrin in Hamsters and Mice." TERATOLOGY 9 (1974): 11-16.

Single oral doses of stated chemicals at 1/2 LD50 to pregnant hamsters caused high incidence of fetal death, congenital anomalies, and growth retardation. This kind of study is the basis of recent action by the Environmental Protection Agency restricting further use of chlorinated hydrocarbon compounds.

477 Petrocelli, Sam R., et al. "Biomagnification of Dieldrin Residues by Food-Chain Transfer from Clams to Blue Crabs under Controlled Conditions." BULLETIN OF ENVIRONMENTAL CONTAMINATION AND TOXICOLOGY 13 (1975): 108-16.

Study of two links in a food chain; both of these animal species are consumed by man. Crabs, fed contaminated clam meat, concentrated dieldrin residues 4.7 to 6.8 times the daily dose after ten days of feeding. No dieldrin was found in water. Authors point out that this was a short-term study and that biomagnification in nature to higher values in tissue is likely. An excellent example of studies needed; good reference list.

478 Porter, Richard D., and Wiemeyer, Stanley N. "Dieldrin and DDT: Effects on Sparrow Hawk Eggshells and Reproduction." SCIENCE 165 (1969): 199-200.

Experimental duplication of effects of pesticidal residues on hawk reproduction in the wild. DDT and dieldrin used. Major effects: increased egg disappearance; increased egg destruction by parent birds; reduced eggshell thickness.

479 Reddy, Gunda, and Khan, M.A.Q. "Fate of Photodieldrin under Various Environmental Conditions." BULLETIN OF ENVIRONMENTAL CONTAMINATION AND TOXICOLOGY 13 (1975): 64-72.

Photodieldrin (PD), the metabolic product of photoaldrin or dieldrin, is considered their "terminal residue." But no one has reported on the fate of PD, its persistence in the environment, or its ecological hazards. This study is preliminary. Residues seem persistent.

480 Scott, Thomas G., et al. "Some Effects of a Field Application of Dieldrin." JOURNAL OF WILDLIFE MANAGEMENT 23 (1959): 409-27.

Disastrous effects on birds and mammals following an application of dieldrin at 3 lbs/acre in efforts to halt the westward spread of Japanese beetle. This report is included here to illustrate the best type of field and laboratory study available for judgment of the effects of environmental pollutants in the 1950s. The concern then has since been shown to be justified. Such field applications are now prohibited.

481 Snyder, Ron L. "Effects of Dieldrin on Homing and Orientation in Deer Mice." JOURNAL OF WILDLIFE MANAGEMENT 38 (1974): 362-64.

Deer mice (Peromyscus maniculatus) are well known for homing ability. Field-taken mice injected with 5 mg/kg dieldrin. Treated animals did poorly in maze orientation and, upon release, did not, in majority of cases, remain in home area. Controls did. Amount injected showed no acutely toxic effects.

482 Stickel, W[illiam]. H., et al. "Tissue Residues of Dieldrin in Relation to Mortality in Birds and Mammals." In CHEMICAL FALLOUT, edited by Morton W. Miller and George G. Berg, pp. 174-204. Springfield, Ill.: Charles C. Thomas, Publisher, 1969.

Good data on residue values from both experimental and wild animals.

483 Walker, A.I.T., et al. "The Toxicology of Dieldrin (HEOD). I. Long-Term Oral Toxicity Studies in Mice." FOOD AND COSMETICS TOXICOLOGY 11 (1972): 415-532.

Two-year study. Mice on low levels in diet (10 ppm and under) showed no signs of ill health. Many pathological changes and deaths at higher levels.

484 Wiemeyer, Stanley N., et al. "Effects of Environmental Pollutants on Connecticut and Maryland Ospreys." JOURNAL OF WILDLIFE MANAGEMENT 39 (1975): 124-39.

Concentrations of residues of dieldrin, PCB, and DDE are higher in Connecticut than in Maryland and seemed to be responsible for the very rapid decline and diminished reproductive success of ospreys in Connecticut. Fish have higher residue levels in New England. Eggshell thickness is 18 percent below normal in Connecticut, 10 percent in Maryland.

485 Winnett, George, and Reed, John P. "Aldrin, Dieldrin, Endrin, and Chlordane Persistence--A 3-Year Study." PESTICIDES MONITORING JOURNAL 2 (1968): 133-36.

Treatments by chemicals were for soil insects; no residues were found in potatoes from treated soil.

BENZENE HEXACHLORIDE (BHC); LINDANE

BHC was developed during World War II and for about twenty years it was commonly used. BHC is a mixture of isomers; when used practically in its pure form, the most toxic is the gamma isomer called lindane. Lindane is still used for special purposes, but BHC as a general insecticide is rarely used now in the United States. Residual persistence is high, approaching that of DDT.

486 Jackson, M.D., et al. "Persistence and Movement of BHC in a Watershed, Mount Mitchell State Park, North Carolina--1967-72." PESTICIDES MONITORING JOURNAL 8 (1975): 202-5.

A good illustration of long-term residue studies. Benzene hexachloride applied at 11.2 kg/ha had residues of 31 and 585 ppm in soil and litter respectively, one month after application. Highest residue in soil was 58 ppm one and one-half years after, declining to 32 ppm five years after application. Remarkable survivability.

487 Tooby, T.E., and Durbin, F.J. "Lindane Residue Accumulation in Rainbow Trout (*Salmo gairdnerii* Richardson) and Roach (*Rutilus rutilis* Linnaeus)." *ENVIRONMENTAL POLLUTION* 8 (1975): 79-89.

Lindane is commonly used in agriculture and forestry in Great Britain. Most waters have measurable background residues of lindane that are in themselves nontoxic. But occasional surges from runoff have caused many fish kills. This study shows, however, that if fish are not killed, both species can eliminate lindane (apparently by hepatic enzymatic induction) in four days or less. There is also no evidence that lindane affects reproduction.

CHLORDANE; HEPTACHLOR; HEPTACHLOR EPOXIDE

Chlordane is the common name of a mixture of compounds with different insecticidal activity. One of the refined compounds is heptachlor, more toxic and persistent than chlordane. However, heptachlor persists even longer, and is more toxic in its environmentally altered form as the epoxide. Normally, it is the epoxide that is found in living tissues. These chemicals are now in little use; termite and ant control are still important urban uses.

488 Andrews, A.K., et al. "Some Effects of Heptachlor on Bluegills (*Lepomis macrochirus*)." *TRANSACTIONS OF THE AMERICAN FISHERIES SOCIETY* 95 (1966): 297-308.

Maximum uptake in tissues occurs within three days. Highest residues within fifty-six days continuous feeding. Fish growth is in inverse proportion to amount of heptachlor. Effects on invertebrates is dose-related.

489 Cochrane, W.P., and Forbes, M.A. "Oxidation Products of Heptachlor and Its Metabolites--A Chemical Study." CHEMOSPHERE 3 (1974): 41-46.

Oxidative products and means of identification by chromatography.

490 Mehrle, Paul M., et al. "Nutritional Effects on Chlordane Toxicity in Rainbow Trout." BULLETIN OF ENVIRONMENTAL CONTAMINATION AND TOXICOLOGY 12 (1974): 513-17.

Authors studied fishes on synthetic diets with 23 or 45 percent protein. Toxicity of chlordane higher in low protein group.

DDT AND ITS DERIVED RELATIVES, DDD AND DDE

DDT epitomizes the entire class of chlorinated hydrocarbon chemicals intentionally applied in pest control. No other synthetic chemical in this book is as widely known throughout the world. No other chemical in pest control has been so liberally applied. And no chemical of concern in environmental toxicology has prompted such a range of investigations and controversy. Literally tens of thousands of articles and many books have been written on the controversial aspects. DDT has the chemical qualities of chlorinated hydrocarbons. These qualities, combined with the hundreds of millions of pounds that have been intentionally released into living systems, ensure continuing problems with its presence even though legislative actions for a decade have successively restricted the manners and amounts of use.

Unintentionally, DDT became the instrument of a great global experiment. Some consequences of that experiment are described in the following annotations. These consequences also dictate a warning not to permit again the release of toxic chemicals with uncontrollable after-effects into the earth's biosphere. The control of DDT and degradation products is not fully effected, but a start has been made. Unintentional pollution from chemicals such as the PCBs, with effects comparable to those of DDT, is more difficult to control. Whatever the arguments regarding DDT's use, purposes, and effects, the development and extensive application of DDT form a standard of comparison by which all pollutants may be judged. DDT now is as much a symbol as it is a chemical compound.

General

491 Beatty, Rita Gray. THE DDT MYTH: TRIUMPH OF THE AMATEURS. New York: John Day Co., 1973. xxii, 201 p.

Beatty, a journalist, challenges environmental activists and DDT use restrictions. Good reading but biased viewpoint.

492 Bitman, Joel, and Cecil, Helene C. "Estrogenic Activity of DDT Analogs and Polychlorinated Biphenyls." JOURNAL OF AGRICULTURAL AND FOOD CHEMISTRY 18 (1970): 1108-12.

There is geometric similarity of these compounds to synthetic estrogen. Both classes of chemicals induce estrogenic activity unless halide or alkyl groups in the p,p' position.

493 Davies, John E., and Edmundson, Walter F., eds. EPIDEMIOLOGY OF DDT. New York: Futura Publishing Co., 1972. x, 157 p.

Fourteen chapters compare exposures and pharmacology of DDT. Noted that DDE is better indicator of long-range exposure, and that exposure to all foreign substances should be considered along with DDT.

494 Dimond, J.B., et al. "DDT Residues in Forest Biota: Further Data." BULLETIN OF ENVIRONMENTAL CONTAMINATION AND TOXICOLOGY 13 (1975): 117-22.

Tables with long-term residue data; for example, ravens had residues to 21.5 ppm total DDT. Dimond's work applies to all treated conifer forests.

495 Fredeen, F.J.H., and Duffy, J. Regis. "Insecticide Residues in Some Components of the St. Laurence River Ecosystem Following Applications of DDD." PESTICIDES MONITORING JOURNAL 3 (1970): 219-26.

Residues measured in water, mud, molluscs, and fish after intentional application in vicinity of Montreal. Most values under 0.5 ppm; one fish had 1.81 ppm in edible flesh.

496 Guenzi, W.D., and Beard, W.I. "Anaerobic Biodegradation of DDT to DDD in Soil." SCIENCE 156 (1967): 1116-17.

No degradation of DDT in sterile soil, but heating and incubation anaerobically resulted in dechlorination (conversion to DDD) by soil microorganisms.

497 Jensen, S., et al. "DDT and PCB in Marine Animals from Swedish Waters." NATURE 224 (1969): 247-50.

Report of pesticide residues in wide range of marine organisms showing marked contamination in the Baltic Sea. Molluscs, fishes, birds, seals were sampled, analyzed for DDT, DDT-R, and PCB. Signs of increase in PCB from north to south. Exceptionally large amounts of residues in white-tailed eagles from Stockholm archipelago. In eagles, values to 25,000 ppm DDT-R in fat; to 14,000 ppm PCB.

498 Kagan, Yu S., et al. "On the Problem of the Harmful Effect of DDT and Its Mechanism of Action." RESIDUE REVIEWS 27 (1969): 43-79.

Main concern is effect on humans but good coverage of mechanisms of action and effect on environment. Extensive literature. Interesting because work comes from the USSR and authors emphasize DDT as a potential menace. This work appeared at about the same time that anti-DDT forces were strongest in the United States.

499 Kuhr, R.J., et al. "DDT Residues in Soil, Water, and Fauna from New York Apple Orchards." PESTICIDES MONITORING JOURNAL 7 (1974): 200-204.

Orchards were heavily treated for a thirteen-year period. DDE and DDT always found in soils. In top six inches, DDT-R ranged from 21.8 to 259 lb/acre. Adjacent stream and aquatic organisms all contained DDT, DDE, and DDD. Highest concentration was in earthworms and slugs.

500 Maugh, Thomas H. II. "DDT: An Unrecognized Source of Polychlorinated Biphenyls." SCIENCE 180 (1973): 578-79.

Reviews particularly contribution of Crosby and Moilanen showing conversion of DDT and metabolites to PCBs. See section II: Polychlorinated Biphenyl Compounds for Work by Crosby and Moilanen.

501 Metcalf, Robert L. "DDT Substitutes." CRITICAL REVIEWS IN ENVIRONMENTAL CONTROL 3 (1972): 25-59.

DDT has given rise to many specific substitutes in insect control; declining use of DDT (because of environmental problems it produces) is paralleled by increasing use of specific chemicals.

502 Woodwell, George M., et al. "DDT in the Biosphere: Where Does It Go?" SCIENCE 174 (1971): 1101-7.

Major review centering on production, dispersal, modeling, and predictions regarding DDT. Essential reading.

503 ______ "DDT Residues in an East Coast Estuary: A Case of Biological Concentration of a Persistent Insecticide." SCIENCE 156 (1967): 821-24.

Excellent, early study establishing the reality of biological accumulation along food chains in complicated ecosystems. Concentrations of DDT increase with trophic level through more than three orders of magnitude, ranging from 0.04 ppm in plankton to 75 ppm in gulls. Highest concentrations occurred in scavenging and carnivorous fish and birds, although birds had 10 to 100 times more than fish. Latter concentrations approach those known to cause death by DDT. Similar concentrations elsewhere in North America.

Soil, Water, and Plants

504 Cox, James, L. "DDT Residues in Marine Phytoplankton." RESIDUE REVIEWS 44 (1972): 23-38.

Transport of residues (notably DDT) into and from the mixed layer of ocean must involve marine phytoplankton and detrital material derived from it. Centrifuged water from California current contains about 2 x 10^{-6} g of DDT residues/g of carbon. Net phytoplankton has about fifteen times as much. Not all organic material concentrates DDT. Good bibliography.

505 ______. "DDT Residues in Marine Phytoplankton: Increase from 1955 to 1969." SCIENCE 170 (1970): 71-73.

Residues of DDT seem to be increasing in the primary stages of coastal pelagic food chains. If processes of decomposition and dispersal of residues are not sufficiently rapid, delay can be expected before decline in domestic use of DDT is reflected in components of food chain.

506 ______. "DDT Residues in Sea Water and Particulate Matter in the California Current System." FISHERY BULLETIN 69 (1971): 443-50.

Concentrations varied from 2.3 x 10^{-12} g/ml off Oregon and Washington to 5.6 x 10^{-12} g/ml off Southern California. Particulate fraction accounted for only 10 percent of residues. Adsorption seems to be uptake mechanism for algal cells. Also supported is idea that 1- to 2-micron diameter particles carry most DDT whether on particles or in water.

507 Dimond, J.B., et al. "Distribution of DDTR in a Uniformly-Treated Stream." BULLETIN OF ENVIRONMENTAL CONTAMINATION AND TOXICOLOGY 12 (1974): 522-28.

Refutes earlier Dimond view that residue values in similarly treated streams are greater in larger than in smaller streams. May not apply to watersheds or where there is more ponding in streams.

508 Giam, C.S., et al. "DDT, DDE, and Polychlorinated Biphenyls in Biota from the Gulf of Mexico and Caribbean Sea--1971." PESTICIDES MONITORING JOURNAL 6 (1972): 139-43.

Levels of DDT, DDE, and PCBs determined in fish, shrimp, crab, and other biota. Residues widely distributed in all biota but coastal water had higher levels than open water.

509 Ginn, Thomas M., and Fisher, Frank M., Jr. "Studies on the Distribution and Flux of Pesticides in Waterways Associated with a Ricefield-Marshland Ecosystem." PESTICIDES MONITORING JOURNAL 8 (1974): 23-32.

Unusually good paper linking biology, analytical chemistry, and "disturbed" environments. Shows aldrin breakdown to dieldrin, prediction of chlorinated hydrocarbon residues for biotic parts of ecosystems, food chain magnification, and concentration of residues in reproductive organs. Most chemicals applied locally, but origins of DDE, DDD, and DDT unknown.

510 Horn, Wilson, et al. "Deposition of DDE and Polychlorinated Biphenyls in Dated Sediments of the Santa Barbara Basin." SCIENCE 184 (1974): 1197-99.

Deposition in dated oceanic sediments began about 1945 for PCB and about 1952 for DDE. Both show progressive increase until 1967. Authors suggest that substantial amounts of chlorinated hydrocarbons are being deposited in deep marine sediments.

511 Peterle, Tony J. "DDT in Antarctic Snow." NATURE 224 (1969): 620.

Antarctic snow melt analyzed showed o,p' and p,p'-DDT. Could be as much as 2.4 x 10^6 kg of DDT in Antarctic snow considering twenty-two years of use and known universal distribution.

512 Smith, Virgil K. "Long-Term Movement of DDT Applied to Soil for Termite Control." PESTICIDES MONITORING JOURNAL 2 (1968): 55-57.

Technical DDT moved very slightly in two decades of weathering in Mississippi.

513 Tarrant, R.F., et al. "DDT Residues in Forest Floor and Soil after Aerial Spraying, Oregon--1965-68." PESTICIDES MONITORING JOURNAL 6 (1972): 65-72.

One month after application of DDT (12 oz/acre), 3 oz/acre of DDT and residues detectable in forest floor. Three years later over 50 percent less. No seeming effect on soil microorganisms.

514 Terriere, Leon C., et al. "Persistence of Pesticides in Orchards and Orchard Soils." In ORGANIC PESTICIDES IN THE ENVIRONMENT, edited by Aaron A. Rosen and H.F. Kraybill, pp. 263-70. Advances in Chemistry Series, no. 60. Washington, D.C.: American Chemical Society, 1966.

In two Oregon orchards examined for DDT and its metabolites, about 50 percent of amount applied could be accounted for. Forty percent of original applied remained as unchanged DDT after twenty years. Most is in the surface layer.

515 Wurster, Charles F., Jr. "DDT Reduces Photosynthesis by Marine Phytoplankton." SCIENCE 159 (1968): 1474-75.

With concentrations as low as a few ppb in water, DDT reduced photosynthesis in laboratory cultures of four species of coastal and oceanic phytoplankton representing four major classes of algae. Toxicity to diatoms increased as cell concentration decreased. This inhibition may be of ecological importance.

Invertebrates

516 Bedford, J.W., and Zabik, M[atthew]. J. "Bioactive Compounds in the Aquatic Environment: Uptake and Loss of DDT and Dieldrin by Freshwater Mussels." ARCHIVES OF ENVIRONMENTAL CONTAMINATION AND TOXICOLOGY 1 (1973): 97-111.

Mussels concentrate DDT about 2,400-fold, dieldrin, 1,200-fold in lake water; about 1,000-fold for DDT in distilled water. Residues highest in digestive and reproductive organs.

517 Burnett, Robin. "DDT Residues: Distribution of Concentrations in Emerita analoga (Stimpson) along Coastal California." SCIENCE 174 (1971): 606-8.

DDT, DDD, and DDE in the beach sand flea reflect sources of contamination. Animals near Los Angeles outfall have forty-five times as much as elsewhere. This concentration is presumed to be due to DDT manufacturing plant effluent. There are probably 100 metric tons of residues in marine sediments off southern California.

518 Butler, P[hilip]. A. "The Significance of DDT Residues in Estuarine Fauna." In CHEMICAL FALLOUT, edited by Morton W. Miller and George G. Berg, pp. 205-20. Springfield, Ill.: Charles C. Thomas, Publisher, 1969.

Good review of monitored and experimental data from invertebrates. Food chain accumulation and extreme acute toxicities in some groups of invertebrates.

519 Cooke, A.S., and Pollard, E. "Shell and Operculum Formation by Immature Roman Snails Helix pomatia L. When Treated with p,p'-DDT." PESTICIDE BIOCHEMISTRY AND PHYSIOLOGY 3 (1973): 230-36.

Relatively low doses of DDT significantly reduced shell operculum weight although high doses did not. After reemergence from hibernation, operculum eating was higher among snails hibernating late in season. As exposure to DDT increased, operculum eating increased. May occur in other species in treated agricultural areas. This phenomenon is comparable to eggshell thinning in birds.

520 Cox, James L. "Uptake, Assimilation, and Loss of DDT Residues by

Euphasia pacifica, a Euphausiid Shrimp." FISHERY BULLETIN 69 (1971): 627-33.

Assimilated amounts of DDT are inversely size-dependent in this abundant marine crustacean. Smaller size classes assimilate more DDT, more rapidly, than larger classes.

521 Dempster, J.P. "The Sublethal Effect of DDT on the Rate of Feeding by the Ground-Beetle Harpalus rufipes." EXPERIMENTAL AND APPLIED ENTOMOLOGY 11 (1968): 51-54.

DDT markedly reduces the feeding activity at concentrations far lower than that required to kill. Recovery is rapid after removal from contact with insecticide. Pest caterpillar survival is therefore enhanced by DDT spraying.

522 Derr, S. Krogh, and Zabik, Matthew J. "Bioactive Compounds in the Aquatic Environment: Studies on the Mode of Uptake of DDE by the Aquatic Midge, Chironomus tentans (DIPTERA: Chironomidae)." ARCHIVES OF ENVIRONMENTAL CONTAMINATION AND TOXICOLOGY 2 (1974): 152-64.

DDE-^{14}C used; no difference in uptake between live or dead fourth instar larvae. Uptake can relate to water hardness. Adsorption-diffusion mechanism proposed to account for mode of uptake and biological concentration capabilities.

523 Odum, W.E., et al. "DDT Residues Absorbed from Organic Detritus by Fiddler Crabs." SCIENCE 164 (1969): 576-77.

DDT and metabolites accumulate in organic plant debris and may persist for years. Residues associated with particulates with diameters of 250-1000 microns. Fiddler crabs, fed detritus with 10 ppm DDT, grossly modified behavior and showed a three-fold increase of DDT residues in muscle of large claw.

524 Soedergren, A., et al. "DDT and PCB in South Sweden." ENVIRONMENTAL POLLUTION 3 (1972): 25-36.

The amphipod Gammurus was used as a prime indicator of contamination but other stream species were also used to determine residue levels, food habits, selective residue values, transport routes, and accumulation rates. Stream organisms need more study.

Fishes and Amphibians

525 Allison, D.T., et al. SOME CHRONIC EFFECTS OF DDT ON CUTTHROAT TROUT. Research Report 64. Washington, D.C.: Fish and Wildlife Service, Bureau of Sports Fisheries and Wildlife, 1964. 28 p.

Several growth and reproductive effects related to amounts of DDT in diet are described. Generally, amounts below 0.1 ppm of DDT seem not to harm fish.

526 Anderson, John M., and Peterson, Margaret R. "DDT: Sublethal Effects on Brook Trout Nervous System." SCIENCE 164 (1969): 440-41.

Suggests that thermal acclimation mechanism is affected by sublethal amounts of DDT. Tail propelling reflex eliminated. Also, DDT prevents visual conditioned avoidance response.

527 Anderson, R.B., and Everhart, W.H. "Concentrations of DDT in Landlocked Salmon (Salmo salar) at Sebago Lake, Maine." TRANSACTIONS OF THE AMERICAN FISHERIES SOCIETY 95 (1966): 160-64.

DDT ranged from 0.87 to 8.8 ppm in salmon organs. Small amounts also found in several other species of fish.

528 Burdick, George E., et al. "The Accumulation of DDT in Lake Trout and the Effect on Reproduction." TRANSACTIONS OF THE AMERICAN FISHERIES SOCIETY 93 (1964): 127-36.

Very important paper. The first to show buildup and storage of DDT in fish flesh and eggs with resultant mortality of young. Phenomenon is comparable to the sequence already shown in birds.

529 Castle, William T., and Woods, Leon A., Jr. "DDT Residues in White Croakers." CALIFORNIA FISH AND GAME 58 (1972): 198-203.

White croakers (Genyonemus lineatus) from coastal Los Angeles area had total DDT residues averaging 10.82 ppm in flesh and 18.23 in flesh plus skin.

530 Cooke, A.S. "The Effects of DDT, Dieldrin, and 2,4-D on Amphibian, Spawn and Tadpoles." ENVIRONMENTAL POLLUTION 3 (1972): 51-68.

Toads are more resistant to either ambient or tissue residues than frogs. Behavioral effects more pronounced with DDT than dieldrin. No effects from 2,4-D.

531 _____. "Selective Predation by Newts on Frog Tadpoles Treated with DDT." NATURE 229 (1971): 275-76.

Warty newts (Triturus cristatus) differentially preyed on tadpoles of the common frog (Rana temporaria) rendered hyperactive by DDT. Author expands on concept of selective predation, impact on prey and predator populations, manner of acquisition of residues by predators, and ultimate alteration of faunal complexes. Weighty ideas in a short paper.

532 Cory, Lawrence, et al. "Distribution Patterns of DDT Residues in the Sierra Nevada Mountains." PESTICIDES MONITORING JOURNAL 3 (1970): 204-10.

Commonest residue in frogs is p,p'DDE at all elevations to 12,000 feet. Highest concentrations were in central and southern regions, declining to north. Highest contamination was on west-facing slope, ascribed to food chain acquisition after DDT aerosol is carried upward from crop dusting in Central Valley. Average values on the west side of the Sierra Nevada were 2-3 ppm, wet weight.

533 Desaiah, D., et al. "DDT: Effect of Continuous Exposure on ATPase Activity in Fish, Pimephales promelas." ARCHIVES OF ENVIRONMENTAL CONTAMINATION AND TOXICOLOGY 3 (1975): 132-41.

Long-term exposure of DDT to fathead minnows in food or water inhibited enzyme activity. Marked in brain and gill tissues.

534 Earnest, Russell D., and Benville, Pete E., Jr. "Correlation of DDT and Lipid Levels for Certain San Francisco Bay Fish." PESTICIDES MONITORING JOURNAL 5 (1971): 235-41.

In 1969, DDT and metabolites were measured monthly in eight species of fish and one species of crab. Significant correlation between percent lipid and DDT-R in most species. Crabs low, perch highest. Detailed tables. Correlates with Lake Michigan studies but contamination is less by factor of 10. It is likely that lesser amounts of DDT residues are available to biota because of higher turbidity and silt adsorption.

535 Giam, C.S., et al. "DDT, DDE and PCB's in the Tissues of Reef Dwelling Groupers (Serranidae) in the Gulf of Mexico and the Grand Bahamas." BULLETIN OF ENVIRONMENTAL CONTAMINATION AND TOXICOLOGY 11 (1974): 189-92.

Pollutant levels generally low. Size of fish is directly related to residue levels. Groupers are not migratory.

536 Malone, Thomas C. "In Vitro Conversion of DDT to DDD by the Intestinal Microflora of the Northern Anchovy, Engraulis mordax." NATURE 227 (1970): 848-49.

Ordinarily DDT dechlorinates rather rapidly to DDE. But in anaerobic conditions DDD can be formed instead. Bacteria and fungi are responsible for conversion to DDD in this experiment, and fungi are responsible for further conversion of DDD to a water-soluble compound.

537 Murphy, Philip G. "The Effect of Size on the Uptake of DDT from

Water by Fish." BULLETIN OF ENVIRONMENTAL CONTAMINATION AND TOXICOLOGY 6 (1971): 20-23.

Size of test fish may be more important determinants of residue loads than the kind of chlorinated hydrocarbon.

538 _____. "Effects of Salinity on Uptake of DDT, DDE and DDD by Fish." BULLETIN OF ENVIRONMENTAL CONTAMINATION AND TOXICOLOGY 5 (1970): 404-7.

DDE constitutes smaller proportion of DDT residues in freshwater fish than in marine fish. Marine fish from estuaries can have DDT residues with 24-35 percent DDE compared with 61-73 percent DDE for the same species from coastal waters. Differential conversions imply that salinity of water affects uptake and metabolism. Reported studies with Gambusia in differing salinities confirm this, but more studies are needed on marine algae and invertebrates.

539 Murphy, Philip G., and Murphy, Josette Vidril. "Correlation between Respiration and Direct Uptake of DDT in the Mosquito Fish Gambusia affinis." BULLETIN OF ENVIRONMENTAL CONTAMINATION AND TOXICOLOGY 6 (1971): 581-88.

Relationship of DDT uptake to body weight is linear, as is O_2 uptake. Surface absorption not important; only live fish pick up DDT. Small fish more efficient in removing DDT from water.

540 Reinert, R[obert]. E. "Pesticide Concentrations in Great Lakes Fish." PESTICIDES MONITORING JOURNAL 3 (1970): 233-40.

Reinert always found DDE, DDD, DDT, and dieldrin in fish. Residue levels varied markedly with species and with size of fish. Experiments show ability of fish to concentrate to ppm level from ppt (parts per trillion) level in water.

541 Reinert, R[obert]. E., and Bergman, H.L. "Residues of DDT in Lake Trout (Salvelinus namaycush) and Coho Salmon (Oncorhyncus kisutch) from the Great Lakes." JOURNAL OF THE FISHERIES RESEARCH BOARD OF CANADA 31 (1974): 191-99.

DDT residues higher in lake trout from southern Lake Michigan in 1966-70 period (average 18.1 ppm) than in Lake Superior in 1968-69 (average 2.2 ppm). Levels increased in both areas with length of fish and amount of oil. Seasonal variation in both oil and residue levels in salmon. Egg concentration relatively high (lake trout, 4.6 ppm; salmon, 7.4-10.2 ppm).

542 Shaw, Stanton B. "DDT Residues in Eight California Marine Fishes." CALIFORNIA FISH AND GAME 58 (1972): 22-26.

DDE, DDD, and DDT present in the livers of all species, from trace amounts to 13 ppm. Amounts in flesh smaller but are in proportion to amounts in livers.

543 Sprague, J.B., and Duffy, J. R[egis]. "DDT Residues in Canadian Atlantic Fishes and Shellfishes in 1967." JOURNAL OF THE FISHERIES RESEARCH BOARD OF CANADA 28 (1971): 59-64.

Variety of marine food items tested for DDT residues. Found lesser amounts off New Brunswick coast than off other North American and North European coasts. Mackerel had highest average values--total DDT residues, 0.54 ppm. Other species (lobster, five species of molluscs, seven fishes) averaged less than 0.1 ppm.

544 Waggoner, John P. III., and Zeeman, Maurice G. "DDT: Short Term Effects on Osmoregulation in Black Surfperch (Embiotoca jacksoni)." BULLETIN OF ENVIRONMENTAL CONTAMINATION AND TOXICOLOGY 13 (1975): 297-300.

Osmoregulatory failure in this marine teleost contaminated with DDT was not sufficient in itself to cause death. Average total DDT residues in fillets of twenty-eight fish were 14.7 ppm.

545 Weis, Peddrick, and Weis, Judith S. "DDT Causes Changes in Activity and Schooling Behavior in Goldfish." ENVIRONMENTAL RESEARCH 7 (1974): 68-74.

Goldfish were exposed to 1.0 ppb in water for seven days. After thirty days they showed significant increases in activity (hyperexcitability) indicated by school's scattering and regrouping slower than controls. Effects not evident after one week in clean water. Ecological significance discussed.

546 Youngs, William D., et al. "Residues of DDT in Lake Trout as a Function of Age." ENVIRONMENTAL SCIENCE AND TECHNOLOGY 6 (1972): 451-52.

In accurately aged trout, residues of p,p'-DDD, p,p'-DDE, and p,p'-DDT increased progressively with age from about 1 ppm at one year to about 14 ppm at twelve years. Correlation between total residues and their ages highly significant.

Mammals

547 Anas, Raymond E. "DDT Plus PCB's in Blubber of Harbor Seals." PESTICIDES MONITORING JOURNAL 8 (1974): 12-14.

Total amounts of both DDT and PCBs were very high, ranging from 2,350 ppm maximum at San Miguel Island, California, to

27.8 ppm maximum in Pribilof Islands, Alaska. Intermediate levels (to 1,620 ppm) in samples from Puget Sound. This species nonmigratory; levels indicate local oceanic pollution.

548 Anas, Raymond E., and Wilson, Alfred J. "Organochlorine Pesticides in Fur Seals." PESTICIDES MONITORING JOURNAL 3 (1970): 198-200.

Thirty samples of liver contained no PCB but all contained DDE; twenty-one contained DDD; twenty-four, DDT; three, dieldrin. In brain tissue, all contained DDE; five, DDD; four, DDT; and none contained dieldrin. DDE was found in three fetuses. Highest value, 5.1 ppm DDE in liver; most values under 1.0 ppm.

549 Anas, R[aymond]. E., and Worlund, Donald D. "Comparison between Two Methods of Subsampling Blubber of Northern Fur Seals for Total DDT Plus PCB's." PESTICIDES MONITORING JOURNAL 8 (1975): 261-62.

Mean DDT plus PCB values were 23.69 ppm in blubber chunks as opposed to 28.64 ppm in same animals but with blubber homogenized.

550 Benson, W.W., et al. "Pesticide and Mercury Levels in Bear." BULLETIN OF ENVIRONMENTAL CONTAMINATION AND TOXICOLOGY 11 (1974): 1-4.

Bear is a game animal, often eaten, in Idaho. Contained lower levels than deer. Highest p,p'-DDE, 2.055 ppb; total DDT, 2.899 ppb. Lowest total DDT, 0.320 ppb. Three bears showed low dieldrin (ppb). Mercury high, 275.0 ppb; low, 40 ppb. Conclude, no hazard.

551 Forsyth, D.J., and Peterle, T[ony]. J. "Accumulation of Chlorine--36 Ring--Labeled DDT Residues in Various Tissues of Two Species of Shrew." ARCHIVES OF ENVIRONMENTAL CONTAMINATION AND TOXICOLOGY 1 (1973): 1-17.

Field study; labeled DDT applied at 0.92 kg/ha. Shrews: Blarina and Sorex. Rapid accumulation and tissue distribution and concentration in shrews. Invertebrates concentrate DDT as food source, to 243 ppm in Blarina fat. Seasonal and sexual effects recorded.

552 Gilbert, Frederick F. "Physiological Effects of Natural DDT Residues and Metabolites on Ranch Mink." JOURNAL OF WILDLIFE MANAGEMENT 33 (1969): 933-43.

Male mink which were fed fish from Miramichi River in New Brunswick (area previously sprayed with DDT) had more residues than females, but effects on spleen and adrenal weights were greater in females. Embryonic losses in minks with DDT-contaminated fish diets were greater than losses in control minks.

553 Hart, Michael M., et al. "Distribution and Effects of DDT in the Pregnant Rabbit." XENOBIOTICA 2 (1972): 567-74.

DDT and metabolites were found in high concentration in fetus and organs regulating fetal environment. Dissipates slowly. At 10 and 50 mg/kg p,p'DDT in diet, much prematurity, fetal resorption, and weight decrease.

554 Sherburne, J[ames]. A., and Dimond, J.B. "DDT Persistence in Wild Hares and Mink." JOURNAL OF WILDLIFE MANAGEMENT 33 (1969): 944-48.

In Maine, after spraying with DDT at 1 lb/acre, adult hares and young show persistent residues over ten years. Residues of 2.9 ppm DDT persist in mink nine years after single application of DDT to forest.

Birds

555 Anderson, Daniel W., et al. "Brown Pelicans: Improved Reproduction off the Southern California Coast." SCIENCE 190 (1975): 806-8.

Productivity up significantly since 1971 when essentially no young were produced (due to high DDT-R loads). Still about thirty percent below normal for population stability, but there is better egg survival and improving eggshell condition. Declining DDE levels in anchovies, pelicans' major food.

556 Bengston, S.A., and Soedergren, A. "DDT and PCB Residues in Airborne Fallout and Animals in Iceland." AMBIO 3 (1974): 84-86.

No residues in air in southwest, low elsewhere. Lower trophic levels in aquatic ecosystems had little or no residues; waterfowl did. Still, levels are one or two orders of magnitude less than those of other reports.

557 Berger, Daniel D., et al. "Shell Thinning in Eggs of Ungava Peregrines." CANADIAN FIELD-NATURALIST 84 (1970): 265-67.

DDE at high levels is 90 percent of the organochlorine residue. Average reduction in eggshell thickness 21 percent. Nest failures occur because of egg breakage and the adults eating shells. Presumes calcium deficiency.

558 Bitman, Joel, et al. "DDT Induces a Decrease in Eggshell Calcium." NATURE 224 (1969): 44-46.

Japanese quail fed 100 ppm of o,p' and p,p'-DDT showed lag

in egg production. Those on p,p'-DDT diet were significantly smaller than control and produced more broken eggs. Shell thickness declined sharply in treated groups. Calcium amounts in treated groups significantly lower. Important paper; first to establish experimental base of eggshell thinning.

559 Blus, Lawrence J., et al. "The Brown Pelican and Certain Environmental Pollutants in Louisiana." BULLETIN OF ENVIRONMENTAL CONTAMINATION AND TOXICOLOGY 13 (1975): 646-55.

This state bird has declined to near extinction. Transplants have been made from Florida. The paper reviews pesticides and heavy metal residue levels as they might relate to the observed decline in reproductive success.

560 _____. "Eggshell Thinning in the Brown Pelican: Implication of DDE." BIOSCIENCE 21 (1971): 1213-15.

Brown pelicans extremely susceptible to DDE effects. PCBs do not contribute significantly to shell thinning.

561 _____. "Further Analysis of the Logarhythmic Relationship of DDE Residues to Eggshell Thinning." NATURE 240 (1972): 164-66.

A rebuttal paper to a challenge regarding the linkage between DDE and shell thinning (see Blus et al., "Logarhythmic Relationship of DDE Residues to Eggshell Thinning," below). Challenge was made by William Hazeltine (NATURE 239 [1972]: 410). This is a good example of the difficulty of presenting correlative data in the presence of strong differences in points of view on the use of DDT.

562 _____. "Logarhythmic Relationship of DDE Residues to Eggshell Thinning." NATURE 235 (1972): 376-77.

Concentration-effect relationship seems to exist between DDE in eggs and shell thinning; follows mathematically similar pattern in different species although it may operate at different levels.

563 _____. "Organochlorine Residues in Brown Pelican Eggs: Relation to Reproductive Success." ENVIRONMENTAL POLLUTION 7 (1974): 81-91.

Eggs were taken from nests in South Carolina colony. Remainder of eggs were watched, and hatching success was judged relative to residue amounts of DDE and dieldrin in sample eggs. Maximum DDE in an egg from a successful nest was 2.37 microg/g; DDE levels were as high as 8.48 microg/g in an unsuccessful nest. The range for dieldrin was 0.54 and 0.99 respectively. Nineteen of twenty-six eggs from unsuccessful nests had residues of DDE exceeding 2.37 microg/g, and fourteen of these twenty-six eggs had residues of dieldrin exceeding 0.54 microg/g. Residues of DDD, DDT, and PCB did not correlate with nest success. Reproductive success of the pelican colony was subnormal in sampling years.

564 _____. "Relation of the Brown Pelican to Certain Environmental Pollutants." PESTICIDES MONITORING JOURNAL 7 (1974): 181-94.

Nearly all brown pelican eggs from thirteen colonies in South Carolina, Florida, and California in 1969 and in South Carolina and Florida in 1970 exhibited eggshell thinning. All eggs had DDE residues; most had measurable quantities of p,p - DDD, p,p'-DDT, dieldrin, or PCBs. All eggs also had some mercury. Although many types of pollutants found, DDE seems to be responsible for most of the thinning. There is strong evidence for this, particularly in South Carolina and in California. Dieldrin may have had some effect in South Carolina. Extensive tabular data in text.

565 Braestrup, Liselotte, et al. "DDE, PCB and Aldrin Levels in Arctic Birds of Greenland." BULLETIN OF ENVIRONMENTAL CONTAMINATION AND TOXICOLOGY 11 (1974): 326-32.

Nine species of birds representing land, shore, and marine forms (with appropriate food habits) were examined. All contained DDE and PCB and many contained lindane. (There is no agriculture or heavy industry in Greenland.) A few high values. For p,p'-DDE: king eiders, 2.8 ppm; guillemot, 8.7 ppm; cormorant, 15.0 ppm; ptarmigan, 4.0 ppm; raven, 18.8 ppm. For PCBs: eiders, 5.3 ppm; guillemot, 39.6 ppm; cormorant, 46.7 ppm; ptarmigan, 15.8 ppm; raven, 63.0 ppm. All values dry weight, neck fat. Particularly high in fish-eating species. Varied literature sources.

566 Brown, N.J., and Brown, A.W.A. "Biological Fate of DDT in a Sub-Arctic Environment." JOURNAL OF WILDLIFE MANAGEMENT 34 (1970): 929-40.

Residues in ecosystems assessed at Fort Churchill, Manitoba, after repeated use of DDT over seventeen-year period. Range of 3 ppm in fat of ptarmigan to 64 ppm in Arctic tern in treated area, compared to 0.2 to 12 ppm in fat of birds from untreated area. There is partial correlation between DDE levels in fat of birds and DDE levels in various food items.

567 Cade, T[om]. J., and Fyfe, Richard. "The North American Peregrine Survey, 1970." CANADIAN FIELD-NATURALIST 84 (1970): 231-45.

Fifteen regions in Canada and Alaska were surveyed for this species, which is declining because of environmental pollutants. Eggshells 15-20 percent thinner and reproductive success lower compared to pre-DDT years. Inverse correlation of chlorinated hydrocarbon residues in eggs and eggshell thickness.

568 Cade, Tom J., et al. "Peregrines and Pesticides in Alaska." CONDOR 70 (1968): 149-53.

Early announcement that pesticide residues, notably DDE, exist in areas far removed from sites of application and that they affect reproduction and behavior due to accumulation in tissues and eggs.

569 Davison, K[enneth]. L., and Sell, J[erry]. L. "DDT Thins Shells of Eggs from Mallard Ducks Maintained on Ad Libitum or Controlled-Feeding Regimes." ARCHIVES OF ENVIRONMENTAL CONTAMINATION AND TOXICOLOGY 2 (1974): 222-32.

Either technical DDT or p,p'-DDT at 20 ppm or more, or dieldrin at 10 ppm caused significant reduction in eggshell thickness, weight, and calcium levels. Thinning correlated linearly with increasing dose of DDT to 40 ppm. DDT at 200 ppm lethal.

570 Dieter, Michael P. "Plasma Enzyme Activities in Coturnix Quail Fed Graded Doses of DDE, Polychlorinated Biphenyl, Malathion and Mercuric Chloride." TOXICOLOGY AND APPLIED PHARMACOLOGY 27 (1974): 86-98.

After twelve weeks on diet, there were increases in enzyme activity of birds on organochlorines and decreases in cholinesterase activity of birds on malathion and mercuric chloride proportional to the dose. After twelve weeks, DDE accumulated fourfold in carcass and liver, but PCB (Arochlor 1254) residues were identical in diet and carcass; PCB in liver was one-half that in diet. Mercury residues did not accumulate more than 5 percent the amount in feed [but mercuric chloride does not accumulate the same way as some other mercury compounds].

571 Dilworth, T.G., et al. "DDE and Eggshell Thickness in New Brunswick Woodcock." JOURNAL OF WILDLIFE MANAGEMENT 36 (1972): 1186-93.

Average thickness of eggshells of woodcock was significantly less in specimens collected in 1971 compared to pre-1925 specimens. Total DDT in eggs was related to amounts used in spruce budworm control.

572 ______. "DDT in New Brunswick Woodcocks." JOURNAL OF WILDLIFE MANAGEMENT 38 (1974): 331-37.

DDT residues were found in muscle of all specimens analyzed

from 1969 to 1971. Average was 2.07 ppm (wet weight), somewhat higher than 1959-61 levels. Amounts in tissue related to forest spraying operations.

573 Dindal, Daniel L. "Accumulation and Excretion of CL^{36} DDT in Mallard and Lesser Scaup Ducks." JOURNAL OF WILDLIFE MANAGEMENT 34 (1970): 74-92.

Single chemical application to marsh at 0.2 lb/acre. DDE commonest metabolite. Higher residues in scaup (animal feeders) than in mallards. Highest residue 67.3 ppm in leg fat of scaup. Long tables; good bibliography.

574 Enderson, James H., and Wrege, Peter H. "DDE Residues and Eggshell Thickness in Prairie Falcons." JOURNAL OF WILDLIFE MANAGEMENT 37 (1973): 476-78.

DDE residues in these birds is lower than in peregrine falcons (because of food habits). But shell thinning notable in 1967-68 eggs, declining in severity in 1972. Reproduction "normal" in 1972.

575 Friend, Milton, and Trainer, Daniel O. "Experimental DDT-Duck Hepatitis Virus Interaction Studies." JOURNAL OF WILDLIFE MANAGEMENT 38 (1974): 887-95.

Mortality and other effects much higher when DDT and virus inoculations are given together rather than singly. Brain tissue levels higher when both DDT and virus given. Conclude DDT residues alter response of ducks to virus.

576 Genelly, Richard E., and Rudd, Robert L. "Effects of DDT, Toxaphene and Dieldrin on Pheasant Reproduction." AUK 73 (1956): 529-39.

One of the first studies of this kind. Pioneering, but rather crude in comparison to later sophisticated studies.

577 Gill, John A., and Verts, B.J. "Tolerances of Two Populations of Ring-necked Pheasants to DDT." JOURNAL OF WILDLIFE MANAGEMENT 34 (1970): 630-36.

Comparison of populations from two areas in Oregon, one with eight-year history of DDT use, the other with no use. Those on untreated areas more tolerant.

578 Gill, John A., et al. "Toxicities of DDE and Some Other Analogs of DDT to Pheasants." JOURNAL OF WILDLIFE MANAGEMENT 34 (1970): 223-26.

LD50 concentrations of p,p'DDT, p,p'DDD, p,p'DDE, and technical (crude formula in normal use) applied to seven-day-old ring-necked pheasants were 550, 522, 1,086, and 935 ppm, respectively. Toxicity of o,p'DDT was in excess of 5,000 ppm. An advantage gained to an individual (by reduced acute toxicity) with exposure to DDE might not be an advantage to a population (longer term effects).

579 Gish, Charles D., and Chura, Nicholas J. "Toxicity of DDT to Japanese Quail as Influenced by Body Weight, Breeding Condition, and Sex." TOXICOLOGY AND APPLIED PHARMACOLOGY 17 (1970): 740-51.

Quail were fed p,p'-DDT for twenty days or until death. Environmental conditions simulated in experiment. Males, nonbreeding birds, and partially starved birds more susceptible to DDT. Cumulative effects described.

580 Greichus, [Y.] A., et al. "Effects of Orally Administered Insecticides and Polychlorinated Biphenyls on the Parasite Fauna of Two Species of Piscivorous Birds." EXPERIMENTAL PARASITOLOGY 35 (1974): 248-52.

Cormorants and white pelicans fed DDT, DDE, DDD, and PCB. Numbers of ectoparasites decreased as dosage of insecticides and residue levels on feathers increased. No difference in endoparasites between treated controls. Also shows that PCBs have insecticidal quality.

581 Haegele, M[ax]. A., and Hudson, Rick H. "DDE Effects on Reproduction of Ring Doves." ENVIRONMENTAL POLLUTION 4 (1973): 53-57.

Ring doves given 40 ppm DDE in diet took an average of 2.5 times longer to renest than controls; produced 13.5 percent fewer eggs per clutch; had 10 percent thinner eggs; and experienced twice as great mortality of young as controls. Rearing success 3.5 times better in control group.

582 _____. "Eggshell Thinning and Residues in Mallards One Year after DDE Exposure." ARCHIVES OF ENVIRONMENTAL CONTAMINATION AND TOXICOLOGY 2 (1974): 356-63.

When under a dietary regime including 40 ppm p,p'-DDE, mallard ducks laid eggs whose shells were 15-20 percent thinner than controls. When DDE was withdrawn, after eleven months, eggshells were still 7.4 percent thinner than controls. At that time tissue residues of DDE were 8.6 ppm.

583 Haegele, Max A., and Tucker, Richard K. "Effects of 15 Common Environmental Pollutants on Eggshell Thickness in Mallards and Coturnix." BULLETIN OF ENVIRONMENTAL CONTAMINATION AND TOXICOLOGY 11 (1974): 98-102.

Most chemicals in diet produced eggshell thinning but the only prolonged effect was from p,p'-DDE in mallard ducks. Temporary eggshell thinning could also be caused by low food consumption.

584 Haseltine, S., et al. "DDE, PTH and Eggshell Thinning in Mallard, Pheasant and Ring Dove." BULLETIN OF ENVIRONMENTAL CONTAMINATION AND TOXICOLOGY 11 (1974): 139-45.

Doves and mallards exhibited thin eggshells and depressed serum calcium after exposure to DDE. PTH (parathyroid hormone) injection increased serum calcium. Pheasants showed neither effects; rather showed long-term thickening under PTH.

585 Haynes, Ronnie J. "Effects of DDT on Glycogen and Lipid Levels in Bobwhites." JOURNAL OF WILDLIFE MANAGEMENT 36 (1972): 518-23.

Quail at 100 ppm DDT in diet showed no significant changes in glycogen or carcass lipids. But liver lipids were significantly higher compared to controls. No conclusion regarding extrapolation to wild birds.

586 Heath, Robert G., et al. "Marked DDE Impairment of Mallard Reproduction in Controlled Studies." NATURE 224 (1969): 47-48.

DDE at 10 and 40 ppm in feed induced significant eggshell thinning and cracking, and increased embryo mortality. DDD and DDT produced less severe effects.

587 Henny, Charles J., et al. "Territorial Behavior, Pesticides and the Population Ecology of Red-shouldered Hawks in Central Maryland, 1943-1971." ECOLOGY 54 (1973): 545-54.

Recruitment rate (number of young fledged per parental pair) was 1.95 in 1940s, but 1.34 after 1960. Change probably due more to habitat modification than to the relatively low pesticide residues present in eggs. Nine percent decline in eggshell thickness in later period. Commonest residue in eggs was p,p'DDE ranging to 1.68 ppm, wet weight. PCBs ranged to 5.0 ppm. Good literature survey.

588 Hunt, L. Barrie. "Physiological Susceptibility of Robins to DDT Poisoning." WILSON BULLETIN 81 (1969): 407-18.

Hundreds of robins died in Madison, Wisconsin, following DDT spraying for Dutch elm disease. Males died sooner and in greater numbers. Seasonal loss of body fat in males an important factor.

589 Hunt, L. Barrie, and Sacho, Roland J. "Response of Robins to DDT and Methoxychlor." JOURNAL OF WILDLIFE MANAGEMENT 33 (1969): 336-45.

Methoxychlor as a substitute for DDT in control of Dutch elm disease produced less mortality in robins. With DDT in use no fledglings were produced; with methoxychlor in use, fledgling success was near normal.

590 Jeffries, D.J., and French, M.C. "Avian Thyroid: Effect of p,p'-DDT in Size and Activity." SCIENCE 166 (1969): 1278-80.

Chemical causes reduction in thyroid weight and collicular colloid content. May be connected with eggshell thinning in birds.

591 _____. "Changes Induced in the Pigeon Thyroid by p,p'-DDT and Dieldrin." JOURNAL OF WILDLIFE MANAGEMENT 36 (1972): 24-30.

Both chemicals induced growth in thyroids and loss of colloids from follicles. Dieldrin had no effect on weights of several organs, but DDE acted much as DDT by increasing liver and adrenal weights and decreasing heart rates. Unexpected mortality suggests that pigeons may be more sensitive to DDE than to DDT.

592 _____. "Hyper- and Hypothyroidism in Pigeons Fed DDT: An Explanation for the 'Thin Eggshell Phenomenon.'" ENVIRONMENTAL POLLUTION 1 (1971): 235-42.

Homing pigeons fed p,p'-DDT showed increase in thyroid weight, reduction in amounts of colloid. Pigeons at low rates of DDT intake were hyperthyroidal; those at high rates, hypothyroidal. Hypothyroidism causes thin shells. Effects are most marked in large birds of prey which are the most highly contaminated with residues.

593 Johnston, David W. "Decline of DDT Residues in Migratory Songbirds." SCIENCE 186 (1974): 841-42.

Analyses of ten species of migratory songbirds killed when birds flew into TV towers in Florida showed a progressive decline of DDT and its metabolites (DDD and DDE) in fat from 1964 to 1973. Decline correlated with decreasing use of DDT during same period.

594 King, Warren B., and Lincer, Jeffrey L. "DDE Residues in the Endangered Hawaiian Dark-rumped Petrel (*Pterodroma phaeopygia sandwichensis*)." CONDOR 75 (1973): 460-61.

Species delining, largely because of mammalian predators. But DDE present to levels of 1.14 ppm in egg and 0.35 ppm in brain. No question that DDT and metabolites measurable in open ocean ecosystem in Central Pacific. For contrast, see 1968 article by C.F. Wurster, Jr., and D.B. Wingate regarding Bermuda petrels, cited in this section.

595 Lehman, J.W., et al. "Effects of DDT on Bobwhite Quail (*Colinus virginianus*) Adrenal Gland." BULLETIN OF ENVIRONMENTAL CONTAMINATION AND TOXICOLOGY 11 (1974): 407-14.

Sublethal amounts increase adrenocortical tissue. There may be subtle endocrinological effects but they are not demonstrated.

596 Linn, Jack D., and Stanley, Ronald L. "TDE Residues in Clear Lake Animals." CALIFORNIA FISH AND GAME 55 (1969): 164-78.

DDD (TDE) accumulation in the biota of Clear Lake is a classic example of biological concentration. Surveys of residues indicate that levels of contamination in birds declined from high in 1958 to much lower in 1963 and 1965.

597 Longcore, J.R., and Samson, F.B. "Eggshell Breakage by Incubating Black Ducks Fed DDE." JOURNAL OF WILDLIFE MANAGEMENT 37 (1973): 390-94.

Fed 10 ppm p,p'-DDE; eggshells thinner 22 percent at equator, 30 percent at cap, and 33 percent at apex than in controls. Natural incubation resulted in fourfold increase in cracking. DDE in eggs averaged 64.9 ppm, wet weight.

598 McLane, M. Anne Ross, and Hall, Linda C. "DDE Thins Screech Owl Eggshells." BULLETIN OF ENVIRONMENTAL CONTAMINATION AND TOXICOLOGY 8 (1972): 65-68.

Owls fed 2.8 ppm DDE in diet through two breeding seasons laid eggs with 12 percent thinner shells. Adds another order (Strigiformes) of birds showing such effects.

599 Mahoney, Joseph J. "Residue Accumulation in White-throated Sparrows Fed DDT for 5 and 11 Weeks." BULLETIN OF ENVIRONMENTAL CONTAMINATION AND TOXICOLOGY 12 (1974): 677-81.

This bird accumulates large amounts of fat in preparation for migration. Fed 5 or 25 ppm; no equilibrium by five weeks; high after eleven weeks. Amounts to 219.4 ppm in brain at 25 ppm in diet. Effects on migration not clear.

600 Miller, D.S., et al. "Enzymatic Basis for DDE-induced Eggshell Thinning in a Sensitive Bird." NATURE 259 (1976): 122-24.

DDE fed to laying Peking ducks. Evidence points toward Ca-ATPase inhibition in the shell gland by DDE, rather than reduced supply of circulating calcium in the blood.

601 Mulhern, Bernard M., and Reichel, William L. "The Effect of Putrefaction of Eggs upon Residue Analysis of DDT and Metabolites." BULLETIN OF ENVIRONMENTAL CONTAMINATION AND TOXICOLOGY 5 (1970): 222-25.

It is often necessary to analyze field-collected eggs that are partially decomposed. This study shows that putrefaction does not affect analysis for DDT.

602 Mulhern, Bernard M., et al. "Organochlorine Residues and Autospy Data from Bald Eagles, 1966-1968." PESTICIDES MONITORING JOURNAL 4 (1970): 141-44.

The presence of PCBs and DDE in all samples (69 specimens) from twenty-five states and the accumulation of high concentrations of dieldrin demonstrates widespread environmental contamination.

603 Peakall, David B. "DDE: Its Presence in Peregrine Eggs in 1948." SCIENCE 183 (1974): 673-74.

Sudden decrease in eggshell thickness of many species of birds began in about 1947, but cause is debated because DDT only came into general use about that year. But peregrine eggs analyzed from 1894 to 1940 showed no residues; those from 1948 to 1950 had residues to 200 ppm in lipids. Alaskan eggs from 1968 had values to 2,000 ppm. (The period 1947-48 can be looked upon as the period when DDT residues became generally incorporated into food chains.)

604 _____. "Pesticide-induced Enzyme Breakdown of Steroids in Birds." NATURE 216 (1967): 505-6.

One of earliest papers linking widespread declines of populations of birds of prey, the occurrence of PCB and pesticide residues, and hepatic enzyme induction as it influences circulation levels of reproductive hormones. Experiments were with DDT and dieldrin in pigeons.

605 Persson, B. "DDT Content of Whitethroats Lower after Summer Stay in Sweden." AMBIO 1 (1972): 34-35.

Whitethroats (warbler: Sylvia communis) caught in spring had significantly higher amounts of DDT and lindane than birds caught in late summer. Implies that birds were subjected to considerably higher contamination from chlorinated hydrocarbons during spring migration through North Africa and Europe than through summer in Sweden, where DDT use has been prohibited since 1970. Difference is sevenfold with PCB; threefold DDT-R; twofold in p,p'DDT.

606 Porter, Richard D., and Wiemeyer, Stanley N. "DDE at Low Dietary Levels Kills Captive American Kestrels." BULLETIN OF ENVIRONMENTAL CONTAMINATION AND TOXICOLOGY 8 (1972): 193-99.

Two birds died after fourteen and sixteen months on diet including 2.8 ppm p,p'-DDE. Brains contained DDE residues of 213 ppm and 301 ppm, compared with average of 14.9 ppm in sacrificed birds showing no prior symptoms.

607 Sladen, William J.L., et al. "DDT Residues in Adelie Penguins and a Crab-eater Seal from Antarctica." NATURE 210 (1966): 670-73.

"This area (Cape Crozier) is as remote from pesticide-contaminated portions of the world as could be hoped for," states Sladen in this article. All specimens had traces of DDT and metabolites. This was the first announcement in detail of this long-distance transport.

608 Soedergren, A., and Ulfstrand, S. "DDT and PCB Relocate When Caged Robins Use Fat Reserves." AMBIO 1 (1972): 36-40.

Fat reserves build up rapidly prior to migration. Metabolism is accelerated; existing organochlorine residues are relocated when fat deposits are mobilized. Experiments involved starvation (to simulate migration) after administration of chemicals. Results show DDT affected retention of PCB and the pattern of retention of PCB in nervous organs. Must consider synergistic effects of different substances.

609 Somers, J., et al. "Effect of External Application of Pesticides to the Fertile Egg on Hatching Success and Early Chick Success and Early Chick Performance. I. Pre-incubation Spraying with DDT and Commercial Mixtures of 2,4-D: Picloram and 2,4-D: 2,4,5-T." BULLETIN OF ENVIRONMENTAL CONTAMINATION AND TOXICOLOGY 11 (1974): 33-38.

Solutions of each chemical were sprayed on eggs (comparable to field applications). No adverse effects. Male chicks hatched from eggs sprayed with DDT and 2,4-D and 2,4,5-T mix gained more weight than controls. Pesticides penetrated shells.

610 Van Velzen, Aldeen [C.], and Kreitzer, J.F. "The Toxicity of p,p'-DDT to the Clapper Rail." JOURNAL OF WILDLIFE MANAGEMENT 39 (1975): 305-9.

Relatively high resistance to DDT in this species. Probably not hazarded by normal marsh treatments with DDT.

611 Van Velzen, Aldeen C., et al. "Lethal Mobilization of DDT by Cowbirds." JOURNAL OF WILDLIFE MANAGEMENT 36 (1972): 733-39.

Birds were fed 100 ppm in diet and then denied almost half their food supply. Authors conclude that stored DDT residues present hazard to birds in times of environmental stress (by utilizing stored fat containing residues).

612 Ware, George W. "Effects of DDT on Reproduction in Higher Animals."

RESIDUE REVIEWS 59 (1975): 119-40.

Good review, notably on eggshell problem; compares PCB effects. Many references.

613 White, Clayton M., et al. "DDE in a Resident Aleutian Island Peregrine Falcon." CONDOR 75 (1973): 306-11.

Migrant prey items (chiefly shorebirds) have 2.8 times as much DDE in tissue as residents. Falcon eggs all had high residues. Desertion of nest common in clutches showing early embryonic death. Island is remote from DDT use.

614 Wurster, Charles F., Jr., and Wingate, David B. "DDT Residues and Declining Populations in the Bermuda Petrel." SCIENCE 159 (1968): 979-81.

Residues of DDT averaging 6.44 ppm in eggs and young of this carnivorous bird indicate widespread contamination of an oceanic food chain remote from applications of DDT.

615 Wurster, C[harles]. F., Jr., et al. "Bird Mortality after Spraying for Dutch Elm Disease with DDT." SCIENCE 148 (1965): 90-91.

Evidence of bird mortality from pesticides difficult to find. A well-studied case illustrating severe effects.

MIREX

Mirex is a near relative to DDT and moves in the environment in similar ways. It was the chemical of choice to replace broadcast heptachlor in attempts to control the imported fire ant in the southern United States. Only in the last few years has it been realized that mirex itself has many undesirable biological effects. Some of these effects are illustrated in the citations to follow. The large-scale program employing mirex has been substantially eliminated by actions of the Environmental Protection Agency. The great persistence and food chain accumulation of mirex are partial reasons for these actions. But also worrisome is the fact that mirex is the most potent inducer of hepatic microsomal enzymes known among the chlorinated hydrocarbons.

616 Borthwick, P.W., et al. "Mirex Residues in Selected Estuaries of South Carolina--June 1972." PESTICIDES MONITORING JOURNAL 7 (1974): 144-45.

Mirex residues in three species of fish: white catfish, 0.021 ppm; bluegill, 0.047 ppm; carp, 0.12 ppm. In blue crabs, 0.026 ppm. None in bottom sediments.

617 Collins, H.L., et al. "Residue Accumulation in Selected Vertebrates

Following a Single Aerial Application of Mirex Bait, Louisiana--1971-72." PESTICIDES MONITORING JOURNAL 8 (1974): 125-30.

Mirex applied at 1.25 lbs of bait per acre. Many species analyzed. Almost all residues below 1.0 ppm. Highest residues (after 3 months) were in mockingbirds and loggerhead shrikes. Food chain accumulation is implied.

618 Davison, K[enneth]. L., et al. "The Effect of Mirex on Reproduction of Japanese Quail and on Characteristics of Eggs from Japanese Quail and Chickens." ARCHIVES OF ENVIRONMENTAL CONTAMINATION AND TOXICOLOGY 3 (1975): 84-95.

Dietary mirex (to 80 ppm) did not affect eggshell weight, fertility, or hatchability in quail. Mirex accumulation in eggs and tissues is proportional to intake and slightly higher in quail than chickens.

619 De la Cruz, A[rmando]. A., and Naqvi, S.M. "Mirex Incorporation in the Environment: Uptake in Aquatic Organisms and Effects on the Rates of Photosynthesis and Respiration." ARCHIVES OF ENVIRONMENTAL CONTAMINATION AND TOXICOLOGY 1 (1973): 255-64.

Striking reduction of photosynthesis in phytoplankton (*Chlamydomonas*) after exposure to 1.0 ppm mirex. Pond snail, *gambusia*, and sunfish initially increase respiratory rate at low concentrations but decrease as much as 50 percent at higher or extended exposures.

620 Hawthorne, Joseph C., et al. "Residues of Mirex and Other Chlorinated Pesticides in Commercially Raised Catfish." BULLETIN OF ENVIRONMENTAL CONTAMINATION AND TOXICOLOGY 11 (1974): 258-64.

No mirex at 0.01 ppm sensitivity level. But all samples contained other pesticide residues. Averages: DDT-R, 1.64 ppm; toxaphene, 1.90 ppm; dieldrin, 0.07 ppm; endrin, 0.05 ppm.

621 Ivie, G. Wayne, et al. "Fate of Mirex--^{14}C in Japanese Quail." BULLETIN OF ENVIRONMENTAL CONTAMINATION AND TOXICOLOGY 11 (1974): 129-35.

Fire ant control agent accumulates in many birds. Male and female quail differ in residue amounts because concentration of mirex residues in egg yolk allows an excretory route for females. Half of residue mirex remained in males after eighty-four days.

622 Jones, Alice Stevens, and Hodges, Charles S. "Persistence of Mirex and Its Effects on Soil Microorganisms." JOURNAL OF AGRICULTURAL AND FOOD CHEMISTRY 22 (1974): 435-39.

Mirex in soil (1:50 ratio) produced no effects on total populations of fungi or bacteria. No degradation of mirex occurred in nine aerobic soils or four anaerobic lake sediments after six-month incubation. Mirex didn't leach from bait or penetrate soil layers after rainfall.

623 Markin, G.P., et al. "Levels of Mirex and Some Other Organochlorine Residues in Seafood from Atlantic and Gulf Coastal States." PESTICIDES MONITORING JOURNAL 7 (1974): 139-43.

Low ranges in crab, oysters, shrimp, and fish. Apparently high residues of mirex which had been reported earlier were partly PCBs.

624 _____. "Residues of the Insecticide Mirex in Terrestrial and Aquatic Invertebrates Following a Single Aerial Application of Mirex Bait, Louisiana--1971-72." PESTICIDES MONITORING JOURNAL 8 (1974): 131-34.

Analyses of twenty-five species of invertebrates showed very little residue in herbivorous forms. Maximum in carnivores and scavengers: spiders to 10.1 ppm in three months; imported fire ants (the target) to 22.1 ppm.

625 Rajanna, Bettaiya, and De la Cruz, Armando A. "Mirex Incorporation in the Environment: Phytotoxicity on Germination, Emergence, and Early Growth of Crop Seedlings." BULLETIN OF ENVIRONMENTAL CONTAMINATION AND TOXICOLOGY 14 (1975): 77-82.

Mirex at values exceeding 0.15 ppm affected all stages indicated in title in several crop species. Mirex is applied at 4.2 g/ha, but amounts in soil are insufficiently known.

626 Spence, James H., and Markin, G.P. "Mirex Residues in the Physical Environment Following a Single Bait Application, 1971-72." PESTICIDES MONITORING JOURNAL 8 (1974): 135-39.

Residues all low (ppb range) in water, soil, and grass, but there is evidence of translocation in plants. Residues peaked at one month in ponds.

627 Tagatz, M.E., et al. "Seasonal Effects of Leached Mirex on Selected Estuarine Animals." ARCHIVES OF ENVIRONMENTAL CONTAMINATION AND TOXICOLOGY 3 (1975): 371-83.

Mirex on fire ant bait (0.3 percent mirex) leached from water in tanks is toxic to blue crabs, pink shrimp, grass shrimp, but not to sheepshead minnows in concentrations less than 0.53 mg/1. Leaching greatest and toxicity highest in summer. All animals concentrated mirex in levels higher than amounts in water: 40,800 x for minnows to 2,300 x for crabs. Sand substrates concentrated 770 x. Field applications of mirex can be toxic to estuarine animals by bioconcentration.

TOXAPHENE (CHLORINATED CAMPHENE)

Toxaphene has been in heavy use as an insecticide for over forty years. It is now more commonly mixed with other classes of insecticides for maximum effectiveness. Early it was discovered that toxaphene had an unusually high toxicity to fish and aquatic invertebrates. Agricultural wastewaters bearing toxaphene have frequently been responsible for fish-kills. Toxaphene has been used as a piscicide in fisheries management, but that use is declining.

628 Casida, John E., et al. "Toxaphene Insecticide: A Complex Biodegradable Mixture." SCIENCE 183 (1974): 520-21.

Composed of at least 175 polychlorinated 10-carbon compounds. Degrades in tissue; survives in environment. About forty million lbs/yr used in the United States, chiefly with methyl parathion on cotton.

629 Chaiyarach, Somchai, et al. "Acute Toxicity of the Insecticides Toxaphene and Carbaryl and the Herbicides Propanil and Molinate to Four Species of Aquatic Organisms." BULLETIN OF ENVIRONMENTAL CONTAMINATION AND TOXICOLOGY 14 (1975): 281-84.

All chemicals widely used in agriculture; runoff or carelessness in application is the problem. Tested on mosquito fish, grass shrimp, crayfish, and a mactrid clam. All species abundant and important in food chains. Toxaphene most toxic to all species, except for molinate's toxicity to clams. Clams are generally most resistant to chemicals. Table of TL_m (median tolerance limits) values.

630 Cushing, C.E., Jr., and Olive, James R. "Effects of Toxaphene and Rotenone on the Macroscopic Bottom Fauna of Two Northern Colorado Reservoirs." TRANSACTIONS OF THE AMERICAN FISHERIES SOCIETY 86 (1957): 294-301.

Toxaphene affected dipteran larvae but not oligochaetes at 0.1 ppm. Rotenone showed less adverse effects at 1.0 ppm.

631 Durant, Charles J., and Reimold, Robert J. "Effects of Estuarine Dredging of Toxaphene-Contaminated Sediments in Terry Creek, Brunswick, Ga.--1971." PESTICIDES MONITORING JOURNAL 6 (1972): 95-96.

Creek is near toxaphene plant outfall. Sediments almost 2,000 ppm; oysters two miles away, 6 ppm.

632 Fukano, F.G., and Hooper, F.F. "Toxaphene (Chlorinated Camphene) as a Selective Fish Poison." PROGRESSIVE FISH CULTURIST 20 (1958): 189-90.

Extreme toxicity to certain species and sizes of fish suggested by fact that to 0.005 ppm toxaphene selectively removes smaller individuals.

633 Hemphill, Jack E. "Toxaphene as a Fish Toxin." PROGRESSIVE FISH CULTURIST 16 (1954): 41-42.

First to note that toxaphene as an insecticide is extremely hazardous to fish, and to suggest its use as a fish poison. Documents fish-kills.

634 Nelson, Judd O., and Matsumura, Fumio. "Separation and Comparative Toxicity of Toxaphene Components." JOURNAL OF AGRICULTURE AND FOOD CHEMISTRY 23 (1975): 984-90.

Fractions show varying toxicities to blue-green algae, brine shrimp, and mosquito larvae. Some are more toxic than general grade toxaphene.

635 Reimold, Robert J., and Durant, Charles J. "Toxaphene Content of Estuarine Fauna and Flora before, during, and after Dredging Toxaphene-contaminated Sediments." PESTICIDES MONITORING JOURNAL 8 (1974): 44-49.

Practical approach to monitoring a stream carrying effluent from a toxaphene manufacturing plant in Georgia. "Impounded" residues showed up in biota during dredging operation. Oysters had toxaphene residues of 2-5 ppm; salt marsh cordgrass (*Spartina*) at high of 7.5 ppm; and a small fish (*Fundulus*) had residues over 200 ppm. Illustrated here are biological concentration of residues; the degree of trapping of residues by sediments; and the effects of disturbing these impounded residues.

Section 6

ORGANOPHOSPHORUS AND CARBAMATE INSECTICIDES

The organophosphorus compounds are all structurally related and there seems to be no limit to the number that can be synthesized. Their history began with the development in World War II Germany of the so-called nerve gases. As a class they are all cholinesterase inhibitors. They interfere with acetylcholine necessary for the transmission of nerve impulses. Their potency depends not only on their intrinsic enzyme affinity but also on anticholinesterase activity acquired through metabolism in the living organism.

This class of compounds has a wide range of solubilities in water and has led, in the case of the most water-soluble kinds, to their use as "systemic" toxicants. These are easily picked up by and translocated in organisms. Most compounds in common use do not depend on this capacity. Although there are many hundreds of these compounds, only some two dozen are commonly used. Perhaps the best known of these are dichlorvos, guthion, malathion, methyl parathion, phosdrin, and phosphamidon. Millions of pounds of these pesticides are applied annually. Their use increased as insect resistance to the chlorinated hydrocarbons grew, and as the serious environmental consequences of the continuing presence of persistent toxicants became fully known. Their increase in use also parallels a period of time in which increasing specificity of pest control action was sought and in which legislation and regulation became increasingly restrictive.

Practical use of organophosphorus compounds is accompanied by many hazards. They are, as a class, highly toxic. Their advantage is that they generally dissipate quickly and do not appreciatively accumulate in plant and animal tissue. Unfortunately, many insect populations have become resistant to them, and no toxicant class is available to substitute for them as was the case for the chlorinated hydrocarbons. Changes in total cultural practices, exemplified in "integrated pest control," appear to be the only reasonable corrective direction to follow.

Carbamate compounds were developed partly in recognition of the developing resistance among insects to organophosphorus compounds. They are not closely related chemically but they are also cholinesterase inhibitors. Only one compound in this class has become a standard insecticide applied in large amounts:

carbaryl or sevin. But several related chemicals--the dithiocarbamates--are used as fungicides.

636 Beynon, K.I., et al. "The Metabolism and Degradation of Vinyl Phosphate Insecticides." RESIDUE REVIEWS 47 (1973): 55-142.

Comprehensive review of chemical groups; includes gardona, phosphamidon, and azodrin. These compounds are lipophilic, relatively toxic, but degrade or are excreted rapidly. Long bibliography.

637 Blau, G.E., and Neely, W. Brock. "Mathematical Model Building with an Application to Determine the Distribution of Dursban[R] Insecticide Added to a Simulated Ecosystem." ADVANCES IN ECOLOGICAL RESEARCH 9 (1975): 133-63.

The model-building process is developed from statistical principles. Authors offer an illustration by building a model describing the fate and distribution of an organophosphate insecticide added to a laboratory system simulating a pond of water.

638 Chemagro Division Research Staff. "Guthion[R] (Azinphosmethyl): Organophosphorus Insecticide." RESIDUE REVIEWS 51 (1974): 123-80.

Summary of chemistry, toxicology, hazards, and formulations of this important insecticide. Toxic to fish but generally low hazard for an organophosphate.

639 CIBA, Ltd. "Phosphamidon." RESIDUE REVIEWS 37 (1971): entire volume.

Ten contributors on several aspects of the chemical. All contributors appear to be associated with CIBA, the Swiss-based chemical firm that developed the pesticide and that is a major worldwide supplier. Contributions include chemistry, toxicology, formulations, application advice, and degradation characteristics.

640 Dorough, H. Wyman. "Metabolism of Insecticidal Methylcarbamates in Animals." JOURNAL OF AGRICULTURAL AND FOOD CHEMISTRY 18 (1970): 1015-22.

Extensive review and reference list. Basically, life of these chemicals in animal bodies is very short, 70-80 percent of the chemicals being eliminated in twenty-four hours.

641 Eto, Morifasa. ORGANOPHOSPHORUS PESTICIDES: ORGANIC AND BIOLOGICAL CHEMISTRY. Cleveland, Ohio: CRC Press, 1974. 387 p.

Technical compendium with 1,096 references.

642 Ferguson, Denzel E., et al. "Toxicity of Dursban to Three Species of Fish." MOSQUITO NEWS 26 (1966): 80-84.

Dursban less toxic than most organochlorine insecticides to shiners, bluegill, and mosquito fish but more toxic than other organophosphate insecticides. Mosquito fish most tolerant. Test fish from insecticide-contaminated environments were three to five times as tolerant as the same species from uncontaminated areas.

643 Iwata, Yutaka, et al. "Behavior of Five Organophosphorus Pesticides in Dust Derived from Several Soil Types." BULLETIN OF ENVIRONMENTAL CONTAMINATION AND TOXICOLOGY 14 (1975): 49-56.

Reason for study is protection of agricultural workers exposed to persistent residues on soil and in dust adhering to fruit and foliage. Many soil types examined. Chemicals, listed in order of decreasing persistence, are ethion, dioxathion, azinphosmethyl, phenthoate, and parathion. All have been implicated in illness of field workers.

644 Kaemmerer, K., and Buntenkoetter, S. "The Problem of Residues in Meat of Edible Domestic Animals after Application or Intake of Organophosphate Esters." RESIDUE REVIEWS 46 (1973): 1-250.

Major review. Veterinary viewpoint. Only a few such chemicals and their actions are well known. Possibly transferable to other animals.

645 Kenaga, Eugene E. "Evaluation of the Safety of Chlorpyrifos to Birds in Areas Treated for Insect Control." RESIDUE REVIEWS 50 (1974): 1-41.

Chemical is active ingredient in dursban, widely used in mosquito and soil insect control. Concludes: no hazard to birds. No mortality or dietary effects. Exemplary study.

646 Kanazawa, Jun. "Uptake and Excretion of Organophosphorus and Carbamate Insecticides by Fresh Water Fish, Motsugo, *Pseudorasbora parva*." BULLETIN OF ENVIRONMENTAL CONTAMINATION AND TOXICOLOGY 14 (1975): 346-52.

This technical paper illustrates two good points: (1) the cholinesterase-inhibiting insecticides supposedly degrade very rapidly in water; (2) there is supposed to be little capacity or opportunity for accumulation in tissues. Neither is precisely true. Fish were exposed at 1.00 ppm in water of each chemical type for thirty days. Malathion was most unstable; fenitrothion moderate (97 percent degraded); diazinon persisted most (72 percent). Among carbamates, carbaryl was most unstable (95 percent in 6 days); BPMC moderate (80 percent in 32 days); XMC most stable (45 percent in 34 days). Uptake more rapid in organophosphates (up to 211 ppm in 3 days with diazinon) but excretion removes most in thirty days. In tanks with diazinon, fenitrothion, and BPMC, deformed fish (with spinal curvature) appeared at frequencies of 10-30 percent.

647 Kuhr, R.J., et al. "Dissipation of Guthion, Sevin, Polyram, Phygon and Systox from Apple Orchard Soil." BULLETIN OF ENVIRONMENTAL CONTAMINATION AND TOXICOLOGY 11 (1974): 224-30.

Report on biodegradability of these organophosphorus or carbamate insecticides. All entered soils but all essentially dissipated by end of growing season. Contrasts with DDT at 235 ppm in top two inches of soil, even though DDT had not been applied for prior ten years.

648 Ludke, J. Larry, et al. "Cholinesterase (CHE) Response and Related Mortality among Birds Fed CHE Inhibitors." ARCHIVES OF ENVIRONMENTAL CONTAMINATION AND TOXICOLOGY 3 (1975): 1-21.

Review of possibility of using CHE inhibition in monitoring programs. Organophosphates increasingly used with decline of chlorinated hydrocarbon insecticides. Generally, 50 percent inhibition produces death. Good literature coverage.

649 Marganian, Vahe M., and Wall, William J., Jr. "Dursban and Diazinon Residues in Biota Following Treatment of Intertidal Plots on Cape Cod--1967-69." PESTICIDES MONITORING JOURNAL 6 (1972): 160-65.

One percent granular dursban at 0.5 lb/acre controlled flies but did not harm crabs or other organisms noticeably. Residues: oligochaetes, 2.30 ppm; mussel, 2.58 ppm; crab, 4.62 ppm; horsefly, 2.58 ppm; snail, 15.7 ppm. Two percent diazinon at 0.20 lb/acre did kill many organisms. Persistence figures for soil and water are given.

650 Menn, Julius J., and McBain, J. Bruce. "New Aspects of Organophosphorus Pesticides. IV. Newer Aspects of the Metabolism of Phosphonate Insecticides." RESIDUE REVIEWS 53 (1974): 35-51.

Phosphonates more toxic than phosphates to mammals and insects. May be because of more efficient intoxication by oxons which are better cholinesterase inhibitors. Good bibliography.

651 Menzer, Robert E., and Dauterman, Walter C. "Metabolism of Some Organophosphorus Insecticides." JOURNAL OF AGRICULTURAL AND

FOOD CHEMISTRY 18 (1970): 1031-37.

Good review of metabolism and "fates" in mammals and insects. Long reference list.

652 Messick, John P., et al. "Aerial Pesticide Applications and Ring-necked Pheasants." JOURNAL OF WILDLIFE MANAGEMENT 38 (1974): 670-85.

Organochlorine and organophosphorus pesticides probably kill some pheasants, especially young birds, but normal use of ethyl parathion, demeton, DDT-toxaphene, and trichlorfon does not cause major losses in southwestern Idaho.

653 Schlagbauer, Birkhild G.L., and Schlagbauer, Albert W.J. "The Metabolism of Carbamate Pesticides--A Literature Analysis, Part I." RESIDUE REVIEWS 42 (1972): 1-84.

Several carbamate insecticides, fungicides, and herbicides reviewed. Important, recently developed chemical group. Warm-bloods degrade residues by hydrolysis; plants and insects by hydroxylation and glycoside formation. Two hundred seventy-nine references. Supplement to this article is cited below.

654 ______. "The Metabolism of Carbamate Pesticides--A Literature Analysis, Part II." RESIDUE REVIEWS 42 (1972): 85-90.

Short supplement to part 1, cited above; extensive review. New literature and information on several compounds.

655 Shure, Donald J. "Insecticide Effects on Early Succession in an Old Field Ecosystem." ECOLOGY 52 (1971): 271-79.

The organophosphate diazinon, applied at 12 lb/acre, was the very highly active ingredient. Plant diversity and biomass increased because chemical differentially affected the weedy plant Convolvulus. There is a trend toward greater insect diversity, apparently because of effects on plants. Factors other than chemical applied, such as soil moisture, influence ecosystem modifications.

656 Silver, Ann. THE BIOLOGY OF CHOLINESTERASES. New York: American Elsevier Publishing Co., 1974. xiv, 596 p.

Definition: cholinesterases are hydrolases which catalyze the hydrolysis of choline esters at a higher rate than that of other esters, and are inhibited at low concentrations of physostigmine and organophosphorous compounds. Relates to our subject by reference to anticholinesterase insecticides, such as parathion. Exceptionally wide breadth, detail, and referencing.

657 Stewart, Paul A. "Summer Decline of Mockingbirds in a South Caro-

lina Agricultural Area." JOURNAL OF WILDLIFE MANAGEMENT 38 (1974): 935-37.

There are few mockingbirds in South Carolina cotton-growing area which is heavily dependent on azodrin for insect control. Contrasts with areas in North Carolina where birds are abundant in agricultural areas. Azodrin is extremely toxic to birds. This article shows how a field biologist's observations lead to speculation about causes for population changes.

658 Waggoner, T. Bill, and Khasawinah, A.M. "New Aspects of Organophosphorus Pesticides. VII. Metabolism, Biochemical, and Biological Aspects of Nemacur and Related Phosphoramidate Compounds." RESIDUE REVIEWS 53 (1974): 79-97.

Applied to soil against chiefly parasitic nematodes. Metabolic pathways in test organisms described. Residues degrade rapidly but do accumulate in tissues.

659 Ward, Diana Valiele, and Howes, Brian Louis. "The Effects of Abate, an Organophosphorous Insecticide, on Marsh Fiddler Crab Populations." BULLETIN OF ENVIRONMENTAL CONTAMINATION AND TOXICOLOGY 12 (1974): 694-97.

Two percent granular formulation in three spaced applications. Might endanger crabs.

660 West, Richard R., et al. "Repelling Pheasants from Sprouting Corn with a Carbamate Insecticide." JOURNAL OF WILDLIFE MANAGEMENT 33 (1969): 216-19.

Seed treatment with a methyl carbamate effectively repelled pheasants. Authors claim ten dollars' worth of corn saved for every dollar of chemical. This claim seems doubtful, particularly because practice certainly has not become common.

PARATHION

661 Benke, G.M., and Murphy, S.D. "Anticholinesterase Action of Methyl Parathion, Parathion and Azinphosmethyl in Mice and Fish: Onset and Recovery of Inhibition." BULLETIN OF ENVIRONMENTAL CONTAMINATION AND TOXICOLOGY 12 (1974): 117-22.

Doses given to sunfish (Lepomis); maximal sublethal symptoms are lethargy, loss of equilibrium, and occasional bursts of hyperactivity. Inhibition much more rapid in mice than fish. In fish, methyl parathion had most rapid effect, then azinphosmethyl and parathion. In mice, azinphosmethyl had most rapid effect, then methyl parathion and parathion. All in all, parathion most toxic to fish. Additive effects likely with continued exposure.

662 Black, W.D., et al. "A Study of the Effect of Ovex on Parathion and Paraoxon Toxicity in Rats." TOXICOLOGY AND APPLIED PHARMACOLOGY 33 (1975): 393-400.

The chlorinated hydrocarbon (Ovex) lessens the toxicity of the chemically unrelated organophosphate (parathion) but does not do so for the more toxic relative (paraoxon). Paper included here to remind that contaminant chemicals should not be thought of singly. Multiple effects can occur and unpredicted consequences are likely with synthetic chemicals.

663 Cook, Sherburne F., Jr., and Conners, Jerrold D. "The Short-Term Side Effects of the Insecticidal Treatment of Clear Lake, Lake County, California, in 1962." ANNALS OF THE ENTOMOLOGICAL SOCIETY OF AMERICA 56 (1964): 819-24.

Control of the Clear Lake gnat was begun with methyl parathion (at 3 ppb) in 1962 after prior treatments with DDD resulted in disastrous biological concentration. With methyl parathion, other than control of the gnat, the only biological effect seemed to be a slight depression of zooplankton populations. (However, in 1975 insecticide resistance in gnats was so high that treatment with methyl parathion in 1976 is not planned.)

664 Hazeltine, William E. "The Development of a New Concept for the Control of the Clear Lake Gnat." JOURNAL OF ECONOMIC ENTOMOLOGY 56 (1963): 621-26.

New concept in 1963 was the substitution of methyl parathion for DDD which had had disastrous accumulative results. Gnats have subsequently developed high resistance, however.

665 Metcalf, Robert L., and March, Ralph B. "Studies of the Mode of Action of Parathion and Its Derivatives and Their Toxicity to Insects." JOURNAL OF ECONOMIC ENTOMOLOGY 42 (1949): 721-28.

Paraoxon, a metabolic product of parathion, is more toxic than its parent chemical and produces toxic effects much more rapidly. Important study. Dates the idea of conversion of an applied chemical to a more toxic "degraded" material.

666 Wolfe, H.R., et al. "Persistence of Parathion in Soil." BULLETIN OF ENVIRONMENTAL CONTAMINATION AND TOXICOLOGY 10 (1973): 1-9.

Parathion generally degrades rather quickly but a small residue persists, adsorbed for long periods. Factors affecting disappearance rate from soil might be adsorption on soil particles, volatilization, leaching, photochemical decomposition, chemical breakdown, and microbial degradation.

667 Yu, Ching-Chieh, and Sanborn, James R. "The Fate of Parathion in a Model Ecosystem." BULLETIN OF ENVIRONMENTAL CONTAMINATION AND TOXICOLOGY 13 (1973): 543-50.

Parathion does not accumulate in trophic sequence, given the conditions of this experiment.

SEVIN

668 Aspoeck, H., and an der Lan, H. "Oekologische Auswirkungen und physiologische Besonderenheiten des Pflanzenschutzmittels Sevin (1-Napthyl-N-methylt carbamate)." [Ecological results and physiological popularities of the plant protectant, sevin] (In German). ZEITSCHRIFT FUER ANGEWANDTE ZOOLOGIE 50 (1963): 343-80.

Good paper; sevin is highly toxic to fish, earthworms, Helicidae, oligochaetes, and Turbellaria, but only slightly toxic to some molluscs.

669 Burdick, George E., et al. "Effect of Sevin upon the Aquatic Environment." NEW YORK FISH AND GAME JOURNAL 7 (1960): 14-25.

A landmark paper, written at the time DDT substitutes were being tested. Sevin at 1.25 lbs/acre in fuel oil resulted in 48.9 to 97.2 percent reduction in invertebrate fish foods in forest streams. Long recovery time. No direct mortality of fish.

670 _____. "Toxicity of Sevin (Carbaryl) to Fingerling Brown Trout." NEW YORK FISH AND GAME JOURNAL 12 (1965): 127-46.

More toxic when applied in aqueous formulation than in oil suspensions. Authors compare three formulations. Toxicity also related to size of fish.

671 Kurtz, David A., and Studholme, Clinton R. "Recovery of Trichlorfon (Dylox^R) and Carbaryl (Sevin^R) in Songbirds Following Spraying of Forest for Gypsy Moth." BULLETIN OF ENVIRONMENTAL CONTAMINATION AND TOXICOLOGY 11 (1974): 78-84.

Analysis following spraying in Pennsylvania at 1.0 lb/acre of trichlorfon and 1.0 lb/acre of sevin. Detectable amounts of trichlorfon in all songbirds after spraying; none in controls. Highest amounts (including metabolite DDVP) in orioles--0.04 ppm. None or trace detected after sevin spray. (Carbaryl is water-soluble and rapidly excreted.)

Section 7
PLANT GROWTH REGULATORS

The targets of herbicides can be roughly divided into selective and nonselective. This distinction derives from the differential effects of herbicides on plant species, and the judgment of what is a desirable or undesirable plant species. Actually, selective herbicides become nonselective if the amount of chemical applied is sufficiently high. Great attention is therefore paid in practice to the formulation--its water solubility and its chemical form. There is no easily identifiable chemical group called herbicides. A great array of types exist. A few are mentioned here because of their common use.

Chlorophenoxy compounds include 2,4-D and 2,4,5-T among many others. The two named are used as selective herbicides. Hundreds of tons of 2,4-D are used annually in aquatic weed control. In general, 2,4-D is directed toward herbaceous species and 2,4,5-T toward woody species. Parent compounds are only slightly water-soluble, the esters are moderately soluble, and the salts completely so. Both compounds are environmentally hazardous, primarily because they alter plant communities. That is their purpose, of course. More recently, the suspicion that 2,4,5-T might be teratogenic has caused some controversy.

Another widely used compound is dalapon, a propionic acid. Dalapon is water-soluble and reasonably stable. Carbamates and dithiocarbamates are toxic to weeds, insects, and fungi. The triazines are widely used on croplands. They vary in water-solubility and stability. The best known is simazine.

Both arsenical and mercurial compounds have been commonly applied. Historically, the inorganic forms were chemicals of choice. Environmental hazard in their use was great. Although organic forms of these two metals are now used primarily, hazard remains high whatever their success as pesticides. The organomercurials have been a particular problem resulting in both controversy and restricted use in many countries. A good example is the banning in some countries of organomercurial compounds on seeds as fungicides. Hazards to human beings, domestic animals, and wildlife have been abundantly documented. See section 10 for references regarding the methylmercury problem.

GENERAL AND MULTIPLE REFERENCES

672 Anderson, A.C., et al. "The Acute Toxicity of MSMA to Blackbass (Micropterus dolomieu), Crayfish (Procambarus sp.) and Channel Catfish (Ictalurus lacustris)." BULLETIN OF ENVIRONMENTAL CONTAMINATION AND TOXICOLOGY 4 (1975): 330-33.

An arsenical herbicide, MSMA (monosodium salt methanearsonic acid) is commonly used along highway rights of way. Laboratory tests with the three species listed in the title indicate good tolerance at rates of field application. However, nothing is known about accumulation in organisms nor about toxicity of degradation products.

673 Ashton, Floyd, and Crafts, Alden S. MODE OF ACTION OF HERBICIDES. New York: Wiley-Interscience, 1973. 504 p.

A major textbook derived from earlier versions by Crafts. Excellent summation.

674 Audus, Lester J., ed. CHEMISTRY AND PHYSIOLOGY. Plant Growth Substances, vol. 1. 3d ed. A Plant Science Monograph. Series ed. Nicholas Polunin. London: Leonard Hill, An Intertext Publisher, 1972. xxiii, 533 p.

Detailed review on plant auxins, both natural and synthetic.

675 Bartsch, Elisabeth. "Diazinon. II. Residues in Plants, Soil, and Water." RESIDUE REVIEWS 51 (1974): 37-68.

Widely used. Degrades rapidly (half-life in 2-4 weeks). Slight translocation in plants. (However, diazinon is quite toxic to animals.)

676 Berry, Charles R., Jr., et al. "Aquatic Macroinvertebrate Response to Field Application of the Combined Herbicides Diquat and Endothall." BULLETIN OF ENVIRONMENTAL CONTAMINATION AND TOXICOLOGY 14 (1975): 374-79.

These chemicals widely used in control of aquatic weeds. In laboratory studies they seem to be quite toxic, particularly to crustaceans. But in field use there seem to be no observable effects at advised levels.

677 Best, Louis B. "First-Year Effects of Sagebrush Control on Two Sparrows." JOURNAL OF WILDLIFE MANAGEMENT 36 (1972): 534-44.

Sagebrush control in Montana reduced sagebrush density, increased grass cover, and reduced forb density. Breeding pairs of Brewer's sparrows were reduced 54 percent following spraying, but no significanct change in vesper sparrows. Study shows the indirect effects of food availability and nest site cover.

678 Cullinore, D.R. "Interactions between Herbicides and Soil Microorganisms." RESIDUE REVIEWS 35 (1971): 65-80.

Literature review of inhibitory effects of herbicides. Other interactions with microorganisms identified. Discusses organic degradation in soils. Bibliography.

679 Edwards, A.C., and Davis, D.E. "Effects of an Organic Arsenical Herbicide on a Salt Marsh Ecosystem." JOURNAL OF ENVIRONMENTAL QUALITY 4 (1975): 215-19.

Technical study of fate of MSMA (Monosodium methane arsonate). Arsenic levels significantly increased in smooth cordgrass, periwinkles, and mussels following applications.

680 Gilderhus, P.A. "Some Effects of Sublethal Concentrations of Sodium Arsenite on Bluegills and the Aquatic Environment." TRANSACTIONS OF THE AMERICAN FISHERIES SOCIETY 95 (1966): 289-96.

At 4.0 ppm, as herbicide, sodium arsenite found to reduce survival, inhibit growth. Immature fish more rapidly affected. Arsenic residues in water, bottom soil, and organs and flesh of bluegills. Bottom fauna and plankton reduced or inhibited.

681 Glass, B.L., and Edwards, W.M. "Picloram and Lysimeter Runoff and Percolation Water." BULLETIN OF ENVIRONMENTAL CONTAMINATION AND TOXICOLOGY 11 (1974): 109-12.

About 0.2 percent picloram (herbicide) appears in deeper percolating waters. Leaching not major route of dissipation.

682 Katan, J., and Eshel, Y. "Interactions between Herbicides and Plant Pathogens." RESIDUE REVIEWS 35 (1973): 145-77.

Herbicides have significant effect on organisms other than targets. Herbicides (from different chemical groups) may increase diseases in crop plants. Some decrease. Many examples; long bibliography.

683 Kearney, Philip C. "Metabolism of Herbicides in Soils." In ORGANIC PESTICIDES IN THE ENVIRONMENT, edited by Aaron A. Rosen and H.F. Kraybill, pp. 250-62. Advances in Chemistry Series, no. 60. Washington, D.C.: American Chemical Society, 1966.

Discusses contribution of soil organisms and enzymes in breakdown of herbicides. Major types described; includes chemical pathways.

684 Kenaga, Eugene E. "Toxicological and Residue Data Useful in the Environmental Safety Evaluation of Dalapon." RESIDUE REVIEWS 53 (1974): 109-51.

Major review of an important herbicide. Dalapon hydrolyzes quickly in water, does not accumulate in food chains, is rapidly degraded in biological systems, and has a low acute and chronic toxicity to wide range of animals. No teratogenicity or significant reproductive effects among birds and mammals tested. Dalapon actually has low toxicity to many plants; specificity is its value. Environmental hazard minimal.

685 Lukens, R.J. CHEMISTRY OF FUNGICIDAL ACTION. New York: Springer-Verlag, 1971. xiii, 136 p.

A balanced source of information on fungicides. Nine chapters.

686 Malone, Charles R. "Effects of a Non-selective Arsenical Herbicide on Plant Biomass and Community Structure in a Fescue Meadow." ECOLOGY 53 (1972): 507-12.

Sodium cacodylate altered community construction and standing crop biomass of vegetation in manner similar to severe heat stress. Repeated applications necessary, resulting in continued simplification of plant community and reduction in biomass during growing season. (Chemical is similar to type used in Viet Nam for defoliation of tropical forests.)

687 Putnam, Alan R., and Penner, Donald. "Pesticides: Interactions in Higher Plants." RESIDUE REVIEWS 50 (1974): 73-110.

Article reviews importance of simultaneous or sequential application of herbicides, fungicides, and insecticides. Cautions that herbicidal use is affected by other chemicals applied. Extensive bibliography.

688 Waid, J.S. "The Possible Importance of Transfer Factors in the Bacterial Degradation of Herbicides in Natural Ecosystems." RESIDUE REVIEWS 44 (1972): 65-71.

Small initial fractions of phenoxy-acetic herbicides (e.g., 2,4-D) survive long periods. Later massive amounts rapidly degraded. Author suggests that genetical mechanisms in microfloras explain difference.

689 Wauchope, R.D. "Fixation of Arsenical Herbicides, Phosphate, and Arsenate in Alluvial Soils." JOURNAL OF ENVIRONMENTAL QUALITY 4 (1975): 355-58.

Organoarsenical herbicides are more strongly sorbed in soils in which phosphates and arsenates are high. Studies of this type useful in predicting environmental fates of some herbicides.

690 Way, J.M. "Toxicity of Hazards to Man, Domestic Animals, and Wildlife from Some Commonly Used Auxin Herbicides." RESIDUE REVIEWS 26 (1969): 37-62.

Widely ranging review of toxicity and hazards to man, laboratory animals, wildlife, bees and other insects, soil, animals, and fish. Herbicides used were MCPA; MCPB; 2,4-D; 2,4-DB; 2,4,5-T; mecoprop; and 2,3,6-TBA. Stresses habitat alterations as well as toxicities.

691 Weaver, Robert J. PLANT GROWTH SUBSTANCES IN AGRICULTURE. San Francisco: W.H. Freeman and Co., 1972. xviii, 594 p.

Advanced textbook and reference covering entire range of endogenous and exogenous (herbicidal) growth substances. Lengthy tables and bibliography. Basal information as well as accounts of "weed-killers."

692 Yu, Ching-Chieh, et al. "Fate of Alachlor and Propachlorin in Model Ecosystem." JOURNAL OF AGRICULTURAL AND FOOD CHEMISTRY 23 (1975): 876-88.

Neither herbicide built up in model food chain ecosystem. Degraded rapidly in water; no metabolites remained in water.

693 _____. "Fate of Pyrazon in a Model Ecosystem." JOURNAL OF AGRICULTURAL AND FOOD CHEMISTRY 23 (1975): 309-11.

Pyrazon is selective herbicide. No evidence of food chain magnification.

2,4-D AND 2,4,5-T

694 Bovey, R.W., et al. "Occurrence of 2,4,5-T and Picloram in Subsurface Water in the Blacklands of Texas." JOURNAL OF ENVIRONMENTAL QUALITY 4 (1975): 103-6.

Both chemicals reached reasonably high levels in surface water shortly after spraying but, in this three-year study, levels in subsurface water were extremely low or nonexistent. Important because these herbicides have been widely used and, at least with 2,4,5-T, the possibility of teratogenicity exists.

695 Dougherty, William J., et al. "The Non-teratogenicity of 2,4,5-Trichlorophenoxyacetic Acid in the Rhesus Monkey (*Macaca mulatta*)." BULLETIN OF ENVIRONMENTAL CONTAMINATION AND TOXICOLOGY 13 (1974): 477-82.

2,4,5-T in oral doses up to 10 mg/kg produced no visible effects in mothers or in young either fetal or to one year of age.

696 Hemmett, Roland B., Jr., and Faust, Samuel D. "Biodegradation Kinetics of 2,4-Dichlorophenoxyacetic Acid by Aquatic Microorganisms." RESIDUE REVIEWS 29 (1969): 191-207.

Experimentally based descriptions. Degradation follows zero-order kinetics. Most important constituent of reaction is the size and nature of microbial populations.

697 Johnson, D.R., and Hansen, R.M. "Effects of Range Treatment with 2,4-D on Rodent Populations." JOURNAL OF WILDLIFE MANAGEMENT 33 (1969): 125-32.

Result of application was increase in grass cover and reduction in herbaceous plants. Density of pocket gophers and chipmunks reduced; voles (*Microtus montanus*) increased in numbers. Changes reflect indirect effect of chemical, since results were due chiefly to changes in type of cover and food availability.

698 Kenaga, Eugene E. "The Evaluation of the Safety of 2,4,5-T to Birds in Areas Treated for Vegetation Control." RESIDUE REVIEWS 59 (1975): 1-20.

Birds should not be affected by residues at any stage in life if recommended dosages followed.

699 Krefting, Laurits W., and Hansen, Henry L. "Increasing Browse for Deer by Aerial Applications of 2,4-D." JOURNAL OF WILDLIFE MANAGEMENT 33 (1969): 784-90.

Clearing undesirable deer foods by use of 2,4-D resulted in greater numbers of deer over eight-year period than in untreated areas.

700 Kopischke, Earl D. "The Effect of 2,4-D and Diesel Fuel on Egg Hatchability." JOURNAL OF WILDLIFE MANAGEMENT 36 (1972): 1353-56.

Simulated natural application with liquid sprays on pheasant eggs. No effect of 2,4-D in commonly used amounts, but diesel oil reduced hatchability to zero.

701 Schultz, Donald P., and Harman, Paul D. "Residues of 2,4-D in Pond Waters, Mud and Fish, 1971." PESTICIDES MONITORING JOURNAL 8 (1974): 173-79.

Much residue data on fate of the dimethylamine salt of 2,4-

dichlorophenoxyacetic acid which is used in large amounts for aquatic weed control. In southern ponds amounts up to 8.96 kg/ha killed no fish, and residues declined within twenty-eight days to "trace only" levels. Effects on reproduction unknown; residues widely distributed in tissues. This study shows that 2,4-D must be considered a common environmental pollutant.

702 Sears, Howard S., and Meehan, William R. "Short-Term Effects of 2,4-D on Aquatic Organisms in the Nakwasina River Watershed, Southeastern Alaska." PESTICIDES MONITORING JOURNAL 5 (1971): 213-17.

Inconclusive in part; no pretreatment data. But seemingly no effects on salmonids or on other aquatic organisms.

Section 8
VERTEBRATE PEST CONTROL

For several reasons the control of vertebrates is not stressed in this volume. Two reasons are paramount. Ordinarily, attempts to control vertebrates are strictly local in scale--a household, a warehouse, a grower's particular acreage, an immediate public health problem. Secondly, although controversy remains, there has not been major change in methodology for decades. The changes that have occurred relatively recently are not extrapolatable to general environmental hazard.

Among the most promising population control techniques are endocrine and gamete regulation. There is a good deal of recent literature on research developments in these methods, but neither is as yet applied in wild populations. If such approaches are applied, one might expect some severe problems in proper limitation and control.

There are currently two major groups of vertebrates in which large-scale efforts to control population numbers are exerted. Selective control of fishes to enhance desired species is widely practiced, and among mammals, area-wide control is still practiced, or at least argued for. Among problems arising from control of fishes are those associated with the chemicals toxaphene and rotenone. For a review of these problems see sections 5 and 12, and R.C. Muirhead-Thompson in section 4.

Among mammals, ground squirrels are now frequently killed by grain baits coated with sodium monofluoroacetate, better known as Compound 1080. Frequently, grains are aerially distributed. The practice continues in parts of the western United States and in New Zealand against other species, including deer and opossum. The technique is hazardous to wildlife and domestic animals because Compound 1080 is highly toxic and because of its stability and its capacity to accumulate in tissue.

The other mammal toward which control efforts have, until recently, been directed on a large scale is the coyote. This fascinating carnivore is charged with great losses to livestock and poultry producers in the western United States. Now banned, but practiced for many years, was area-wide placement of flesh-bait stations by governmental agencies. Baits were injected with Compound

1080. The hazard to flesh-eating birds and mammals either by direct or secondary poisoning was enormous. Another method of controlling coyotes is the so-called coyote-getter. This device, placed in flesh baits, has a propellant which shoots sodium cyanide into the mouth of the predator. It is somewhat more selective than the 1080-flesh stations, but it does kill dogs and other mammals. Although banned for some years, the method is back in limited practice in California and Texas on an experimental basis.

No attempts to control birds and mammals can pass without emotional and legislative contention. Judgments on degree of hazard from these control methods are clouded, even in rational people, by differing systems of value.

703 Bamford, John, and Martin, John. "A Method for Predicting Success of of Aerial Poison Campaigns against Opossums." NEW ZEALAND JOURNAL OF SCIENCE 14 (1971): 313-21.

Methods of predicting kill success are based solely on weights and lengths of opossums. Measurements are derived from control operations against opossums using 1080--impregnated pellets.

704 Brooks, Joe E. "A Review of Commensal Rodents and Their Control." CRITICAL REVIEWS IN ENVIRONMENTAL CONTROL 3 (1973): 405-53.

Generally interesting review; included here because of its summary presentations of most rodenticides.

705 Janda, Jiri, and Boesseova, Marie. "The Toxic Effect of Zinc Phosphide Baits on Partridges and Pheasants." JOURNAL OF WILDLIFE MANAGEMENT 34 (1970): 220-23.

Zinc phosphide at 2.5 percent on wheat grains was persistent in moist air and highly poisonous to game birds. Intended targets were small rodents.

706 Robinson, William H. "Acute Toxicity of Sodium Monofluoroacetate to Cattle." JOURNAL OF WILDLIFE MANAGEMENT 34 (1970): 647-48.

LD50 for sodium monofluoroacetate (Compound 1080) is 0.221 mg/kg. That is extremely toxic; rodent baits using the compound should be applied sparingly.

707 Rudd, Robert L. PESTICIDES AND THE LIVING LANDSCAPE. Madison: University of Wisconsin Press, 1964. xi, 320 p.

Includes both technical and social discussion of vertebrate pest control programs. Also cited in section 4: Pesticides.

708 Rudd, Robert L., and Genelly, Richard E. PESTICIDES: THEIR USE AND TOXICITY IN RELATION TO WILDLIFE. California Department

of Fish and Game, Game Bulletin no. 7. Sacramento: State Printing Office, 1956. 209 p.

Reviews vertebrate poisons. Also cited in section 4: Pesticides.

709 Terriere, Leon C., and Burnard, R.J. "Uptake, Tissue Distribution, and Clearance of the Selective Piscicide 1,1'-Methylenedi-2-Naphthol (Squoxin) by the Rainbow Trout and the Squawfish." JOURNAL OF AGRICULTURAL AND FOOD CHEMISTRY 23 (1975): 714-17.

This phenol compound selectively toxic to squawfish at concentrations not damaging to trout. Trout about one-twentieth as sensitive.

710 von Oettingen, W.F. THE TOXICITY AND POTENTIAL DANGERS OF ZINC PHOSPHIDE AND OF HYDROGEN PHOSPHIDE (PHOSPHINE). (A REVIEW OF THE NONSECRET, NONCONFIDENTIAL AND NONRESTRICTED LITERATURE). Public Health Reports, Supplement no. 203. Washington, D.C.: Government Printing Office for Public Health Service, 1947. 17 p.

The only good review of effects of zinc phosphide.

Section 9

SEX ATTRACTANTS AND CHEMOSTERILANTS

The problems created by broad-spectrum insecticides have led to increased study and support of methods that might be both pest-specific and applicable on a large scale. Among the most interesting of these methods under active study is the use of specific chemicals produced by the pest species itself. Pheromones are chemical substances which are secreted by an organism and act at a distance from the individual to influence another individual of the same species. Sex pheromones are one such group of substances. The approach may be generally useful in time, and much effort currently goes into research on the subject. The best that can be said at this time is that female sex hormone attractants in gypsy moth have been successfully used in monitoring populations of that species; and on an experimental basis, limited population control of the Western pine beetle has been successful. But the possibilities for use are great and desirable.

Some background references for understanding this potential substitute for toxic substances follow. Mention is also made of insect chemosterilants, not yet applied in practice. Although pheromones would seem to pose no general environmental hazard, chemosterilants conceivably could. Much would depend upon the degree of species specificity and the nature of the formulation applied.

711 Beroza, Morton, ed. CHEMICALS CONTROLLING INSECT BEHAVIOR. New York: Academic Press, 1970. xvi, 170 p.

Contains six contributed papers on pheromones, attractants, and repellents. Practical and scientific considerations.

712 Birch, Martin C., ed. PHEROMONES. New York: American Elsevier Publishing Co., 1974. xxi, 495 p.

Twenty-nine contributors in twenty-two chapters. Chiefly insects, but four chapters on vertebrates, including man. Concluding two chapters on population management with pheromones.

713 Jacobsen, Martin. "Insect Pheromones." In THE PHYSIOLOGY OF INSECTS, edited by Morris Rockstein, vol. 3, pp. 229-76. New York: Academic Press, 1974.

Excellent review and literature list on chemical communication in insects. Sex attractants (basis for some insect control methods) are described.

714 Labrecque, Germain C., and Smith, Carroll N., eds. PRINCIPLES OF INSECT CHEMOSTERILIZATION. New York: Appleton-Century-Crofts, 1968. viii, 354 p.

Seven contributors; excellent review of the chemistry, toxicology, and applied aspects of insect chemosterilization.

715 Lanier, G.N., and Wood D[avid]. L. "Specificity of Response to Pheromones in the Genus Ips (COLEOPTERA: SCOLYTIDAE)." JOURNAL OF CHEMICAL ECOLOGY 1 (1975): 9-23.

An experimental study of seventeen species of Ips (an important pest in forests) to determine whether pheromones are species-specific in inducing aggregations. This 1975 article shows that work is proceeding with alternatives to insect control by broadcast toxic chemicals. Compare with gypsy moth sex attractants.

716 Marx, Jean L. "Insect Control: Use of Pheromones." SCIENCE 181 (1973): 736-37.

Sex attractants have had notable success in the gypsy moth detection program and have experimentally been responsible for great reductions in Western pine beetle populations. Generally, though, pheromonal control is chiefly in the theoretical stage, whatever its desirability as a replacement for broad-spectrum pesticides.

717 Schooley, David A., et al. "Environmental Degradation of the Insect Growth Regulator Methoprene (Isopropyl [2E,4E]-11-Methoxy-3,7,11-Trimethyl-2,4-Dodecadienoate). II. Metabolism by Aquatic Organisms." JOURNAL OF AGRICULTURE AND FOOD CHEMISTRY 23 (1975): 293-98.

First insect growth regulator to be approved by Environmental Protection Agency as a commercial mosquito larvicide. Has half-life of about forty hours in pond water. See also "III. Photodecomposition," by these authors in the same journal issue, pages 299-303.

718 Sheads, R.E., et al. "Chemicals Related to the Gypsy Moth Sex Pheromone as Attractants and as Synergists or Inhibitors of the Pheromone."

JOURNAL OF AGRICULTURAL AND FOOD CHEMISTRY 23 (1975): 60-64.

Eighteen chemicals related to the natural product (called "disparlure") were neither as effective nor as synergistic.

719 Wood, David L., et al, eds. CONTROL OF INSECT BEHAVIOR BY NATURAL PRODUCTS. New York: Academic Press, 1970. x, 345 p.

Nineteen contributed papers, chiefly about pheromones.

Part III

CONSEQUENCES OF UNINTENDED ENVIRONMENTAL POLLUTION

Intentional pollution with toxicants has produced effects in places that may be serious, and there has been widespread environmental contamination because of intended use. But such forms of intentional contamination can be controlled in part simply by discontinuing the use, although this is not always possible in a relatively short time span. But unintended pollutants cannot be treated that way as easily, and they can be considered potentially more hazardous simply because they often contaminate on a grand scale before there is an awareness of their presence or consequences. The problem is simpler when a "point" source such as a smelter or a manufacturing plant is determined as the source of pollution: correctives can be taken, albeit by regulation and with contention. But the solution becomes incredibly difficult when the waste products of our entire industrial culture are involved.

Sections 10 and 11 deal with two classes of chemicals that, from our standpoint, can only be properly described as the wastes of man's own product invention. Section 12 deals with natural toxic inventions among plant and animals. These biological substances usually involve getting something to eat or avoiding being eaten. There are thousands of such chemicals, and many varied morphological adaptations of these chemicals among living things.

Section 10

METALS

Metals and their compounds do not naturally occur uniformly in nature whether in earth or in living tissue. Some areas are known to have unusually high or low amounts of particular compounds. Areas of high selenium content have been identified by their effects on grazing livestock; cobalt deficiency has been identified in Australian sheep. Most metals are necessary as traces for biotic maintenance. Too little or too much is therefore not an unexpected consequence of the mosaic distribution of naturally occurring metals.

Man's use of metals also results in uneven distribution. They are usually smelted at a point near a source of ore, with subsequent processing at a point distant from both their natural occurrence and the point of their first production. In spite of all these steps, their occurrence is at a reasonably describable place or point. Any adverse effects on the biota begin with a point source and proceed in a gradient of decreasing intensity to another point at which no adverse effects occur. The pattern is not one of universal contamination but of focal, source-gradient effects. A chlor-alkali plant situated on a stream, for example, produces gradually decreasing effects with its mercury-containing effluents downstream, and none upstream. Point-source effects may be environmentally serious and they may be subject to controversy and economic disruption, but they are controllable in time.

The more serious, only secondarily controllable, problem is that man is an agent in the redistribution of metals, toward uniform occurrence throughout the earth. Change from mosaic to uniform distributions, whether of metals, DDT, or atomic radiation, is a major concern to environmentalists. Metals can be converted into organic compounds either naturally or by man. The uncontrolled and unpredictable consequences of more uniformly distributed waste products converting to highly toxic organo-metallic compounds is a frightening prospect. The conversion of inorganic mercury to methyl mercury is an illustration.

Many difficult problems remain to be resolved with the industrial use of metals and their redistribution. The following articles describe some of the areas of study. General and multiple references are listed first, followed by six metals separately identified. These six are presently causing serious environmental hazards.

GENERAL AND MULTIPLE REFERENCES

720 Anderlini, Victor. "The Distribution of Heavy Metals in the Red Abalone Haliotis rufescens, on the California Coast." ARCHIVES OF ENVIRONMENTAL CONTAMINATION AND TOXICOLOGY 2 (1974): 253-66.

Analysis for eight heavy metals (Ag, Cd, Cr, Cu, Pb, Hg, Ni, and Zn). High cadmium (1400 ppm) in digestive gland presumed to be background normal. Copper and silver concentrations inversely correlated north to south gradient. High mercury content off southern California; elsewhere mercury seemingly naturally occurring.

721 Anderson, J.W., et al. "Sublethal Effects of Oil, Heavy Metals and PCB's on Marine Organisms." In SURVIVAL IN TOXIC ENVIRONMENTS, edited by M.A.Q. Khan and John B. Bederka, Jr., pp. 83-121. New York: Academic Press, 1974.

General problem is comprehending effects of increasing pollution in near-shore and estuarine environments. Detailed article; table 1 gives an extended outline of the entire range of effects.

722 Baudouin, M.F., and Scoppa, P. "Acute Toxicity of Various Metals to Freshwater Zooplankton." BULLETIN OF ENVIRONMENTAL CONTAMINATION AND TOXICOLOGY 12 (1974): 745-51.

Industrial pollution in relation to food of fishes. Three genera (Cyclops; Eudiaptomus; Daphnia) were employed but authors decided that the broadly available Daphnia was best indicator and experimental animal. Extensive tables and experimental procedures. Toxicities (48h-LC50) of twelve metals given. Mercury, copper, and zinc most toxic.

723 Calabrese, Anthony, and Nelson, David A. "Inhibition of Embryonic Development of the Hard Clam, Mercenaria mercenaria, by Heavy Metals." BULLETIN OF ENVIRONMENTAL CONTAMINATION AND TOXICOLOGY 11 (1974): 92-97.

Metals, as inorganic salts: mercury, silver, zinc, nickel, lead. Laboratory animals were used, with artificial seawater as medium to standardize. Mercury and silver most toxic. Mercury 100 percent lethal at 0.0075 ppm; silver at 0.045 ppm. Zinc 100 percent lethal to clam embryoes at 0.25 ppm.

724 Duce, R.A., et al. "Enrichment of Heavy Metals and Organic Compounds in the Surface Microlayer of Narraganset Bay, Rhode Island." SCIENCE 176 (1972): 161-63.

Concentrations of lead, iron, nickel, copper, fatty acids, hydrocarbons, and chlorinated hydrocarbons enriched 1.5 to 50 times in top 100 to 150 micrometers of surface water, relative to bulk water 20 meters below surface. Produces high pollutant concentration in surface microlayer. Regular source may be atmospheric.

725 Eisler, Ronald. ANNOTATED BIBLIOGRAPHY ON BIOLOGICAL EFFECTS OF METALS IN AQUATIC ENVIRONMENTS (no. 1-567). Environmental Protection Agency, EPA-R3-73-007. Washington, D.C.: Government Printing Office, 1973. 287 p.

Total of 567 references annotated and indexed by metal and taxa. Preference to heavy metals in marine, estuarine, and anadromous species. Good analyses.

726 Environmental Protection Agency. CYCLING AND CONTROL OF METALS. Cincinnati, Ohio: National Environmental Research Center, 1973. vi, 187 p.

Variety of aspects from chemistry to regulation of trace and heavy metals as pollutants. Many contributors to volume.

727 Gibbs, Robert H., Jr., et al. "Heavy Metal Concentrations in Museum Fish Specimens: Effects of Preservatives and Time." SCIENCE 184 (1974): 475-77.

Specimens of myctophid fish (lantern fish) preserved for one month in formalin, ethyl alcohol, or isopropyl alcohol had higher concentrations of cadmium, copper, zinc, and sometimes lead, and lower concentrations of mercury, and sometimes lead, than did unpreserved frozen specimens. Comparisons of concentrations of metals must take manner of preservation into account.

728 Graham, David L. "Trace Metals in Intertidal Molluscs of California." VELIGER 14 (1972): 365-72.

Seven species of molluscs from eleven California coastal stations were analyzed for contents of silver, cadmium, chromium, copper, manganese, lead, and zinc. Provides baseline data (natural levels) against which pollutant intrusions may be judged. Literature cited in this article is a good source of similar data.

729 Gupta, R. Sen. "On Some Trace Metals in the Baltic." AMBIO 1 (1972): 226-30.

Basal study of water quality; zinc, copper, nickel, cobalt, iron, and lead appraised. Iron and copper concentrations fair assessors of quality in anoxic zones. Chemistry of nickel in seawater discussed.

730 Iskandar, Iskandar K., and Keeney, Dennis R. "Concentration of Heavy Metals in Sediment Cores from Selected Wisconsin Lakes." ENVIRONMENTAL SCIENCE AND TECHNOLOGY 8 (1974): 165-70.

Concentration of heavy metals (cadmium, lead, and others) in sediment cores. Precultural levels much lower than postcultural, particularly modern (0-10 cm). Copper related to algacidal use; lead attributable to atmospheric loading from engines.

731 Knauer, George A., and Martin, John H. "Seasonal Variation of Cadmium, Copper, Manganese, Lead, and Zinc in Water and Phytoplankton in Monterey Bay, California." LIMNOLOGY AND OCEANOGRAPHY 18 (1973): 597-604.

Good environmental study of metals distribution along transect from coast to far offshore.

732 Murthy, G.K. "Trace Elements in Milk." CRITICAL REVIEWS IN ENVIRONMENTAL CONTROL 4 (1974): 1-37.

Study of trace metals and radionuclides in cow's milk. Compare with synthetic organic compounds in milk.

733 National Research Council. VANADIUM. Washington, D.C.: National Academy of Sciences, 1974. vii, 117 p.

Contribution from a panel of specialists, part of the Committee on Biological Effects of Atmospheric Pollutants. Vanadium is an actual problem in metal alloying and in oil refining. But it occurs in plant, animal, and human tissue. Possibility that widespread toxicity exists.

734 Nicholas, D.J.D., and Egan, Adrian R., eds. TRACE ELEMENTS IN SOIL-PLANT-ANIMAL SYSTEMS. New York: Academic Press, 1975. xiv, 417 p.

Twenty-three contributors considering wide range of metals and effects. Among others, selenium, cobalt, boron, zinc, manganese, and copper are mentioned.

735 Penrose, W.R. "Arsenic in the Marine and Aquatic Environments: Analysis, Occurrence, and Significance." CRITICAL REVIEWS IN ENVIRONMENTAL CONTROL 4 (1974): 465-86.

Good background article on "the oldest poison." Arsenic occurs naturally in marine environments; arsenical compounds are sometimes used in pest control. Whichever source, amounts in food cannot exceed 2.6 ppm (U.S. regulation) or 1.0 ppm (Great Britain regulation).

736 Preston, A., et al. "British Coastal Waters: The Concentration of Selected Heavy Metals in Sea Water, Suspended Matter and Biological Indicators--A Pilot Survey." ENVIRONMENTAL POLLUTION 3 (1972): 69-82.

Useful summary of heavy metal levels in coastal and pelagic waters around Great Britain. Concentrations are highest in the Irish Sea. Levels have changed little in the decade 1960-70.

737 Schroeder, Henry A. THE POISONS AROUND US. TOXIC METALS IN FOOD, AIR, AND WATER. Bloomington: Indiana University Press, 1974. xii, 124 p. Tables.

Rather curious approach--popularized and a bit facile--to the toxic trace metals, primarily cadmium, beryllium, antimony, mercury, and lead. Contains ten chapters beginning with "Too Much Is Too Much" and ending with "Control of Metal Pollution."

738 Skidaway Institute of Oceanography. TRANSPORT, FATE AND GEOCHEMICAL INTERACTIONS OF MERCURY, CADMIUM AND OTHER INORGANIC POLLUTANTS IN THE COASTAL LITTORAL--SALT MARSH ENVIRONMENT OF THE SOUTHEASTERN UNITED STATES. Springfield, Va.: National Technical Information Service, U.S. Dept. of Commerce, 1974. 242 p. Figures and tables.

Report to Environmental Protection Agency. Good statement of techniques and problems of this important, regularly polluted environment.

739 Tyler, Germund. "Heavy Metals Pollute Nature, May Reduce Productivity." AMBIO 1 (1972): 52-59.

General article centering on Sweden. Increasing deposition of heavy metals due to human activity may reduce the decomposition rate, effecting decrease in ecosystemic productivity. Forest fires or clear cutting might suddenly release deposited heavy metals.

740 Zitko, V., et al. "Thallium: Occurrence in the Environment and Toxicity to Fish." BULLETIN OF ENVIRONMENTAL CONTAMINATION AND TOXICOLOGY 13 (1975): 23-30.

Small amounts of thallium are present in ores of the base metals but commercial demand is too small to effect recovery. Therefore, this highly toxic metal may be found in effluents from smelting operations. Its toxicity to Atlantic salmon is about the same as copper's. The environmental problem is local, limited to areas of mineral processing. Study in New Brunswick. Good references.

CADMIUM

741 Bengtsson, B[engt].-E[rik]., et al. "Vertebral Damage in Minnows Phoxinus phoxinus L., Exposed to Cadmium." AMBIO 4 (1975): 166-68.

Surviving fish (31 of 101) exposed to sublethal amounts of cadmium for seventy days showed lesions in spinal column. Pollutant students must take subtle exposures into account.

742 Cearley, Jack E., and Colman, Ronald L. "Cadium Toxicity and Bioconcentration in Largemouth Bass and Bluegill." BULLETIN OF ENVIRONMENTAL CONTAMINATION AND TOXICOLOGY 11 (1974): 146-51.

Cadmium at 0.85 mg/1. Fishes concentrated cadmium to level higher in the water, bass more than bluegill. Some damage to nervous systems of both species suggested by aberrant behavior.

743 Clubb, Robert W., et al. "Acute Cadmium Toxicity Studies upon Nine Species of Aquatic Insects." ENVIRONMENTAL RESEARCH 9 (1975): 332-41.

Good study of its type. Only mayflies and stoneflies sensitive to cadmium. [Relate this study to water pollution caused by increasing use of cadmium in manufacturing and in effluents from factories.]

744 _____. "Synergism between Dissolved Oxygen and Cadmium Toxicity in Five Species of Aquatic Insects." ENVIRONMENTAL RESEARCH 9 (1975): 285-89.

Toxicity of cadmium increases as dissolved oxygen increases. Amount in insect tissue also increases. (Toxicity of a compound may vary with environmental conditions.)

745 Klass, Evan, et al. "The Effect of Cadmium on Population Growth of the Green Alga Scenedesmus quadracauda." BULLETIN OF ENVIRONMENTAL CONTAMINATION AND TOXICOLOGY 12 (1974): 442-45.

Authors find that cadmium does affect population growth. The 10 ppb guideline established as "safe" for human consumption should be lowered to protect all members of the aquatic community.

746 Martin, M.H., and Coughtrey, P.J. "Preliminary Observations on the Levels of Cadmium in a Contaminated Environment." CHEMOSPHERE 4 (1975): 155-60.

Levels of cadmium in a woodland ecosystem in England are described. Cadmium accumulates in the surface (more organic) layers of soil and in mosses. Tables detail both taxonomic and trophic level concentrations. Food chain accumulation exists. Some invertebrate herbivores (earthworms, isopods) contain very high levels as do vertebrate carnivores (song thrushes, sparrow hawks). There are few ecosystem studies of this kind with heavy metals involved.

747 Miller, David W., et al. "The Uptake and Distribution of Cadmium--115 m in Calcium Deficient and Zinc Deficient Golden Hamsters." BULLETIN OF ENVIRONMENTAL CONTAMINATION AND TOXICOLOGY 13 (1975): 40-43.

Straight toxicological study. The problem is that cadmium replaces the necessary elements calcium and zinc to exert its toxic effects.

748 Nielsen, Stephen A. "Cadmium in New Zealand Dredge Oysters: Geographic Distribution." INTERNATIONAL JOURNAL OF ENVIRONMENTAL AND ANALYTICAL CHEMISTRY 4 (1975): 1-7.

High cadmium levels--to 9 ppm, wet weight; must be naturally occurring, for there is no industrial source. Perhaps *Ostrea* differentially (compared with other forms) concentrates cadmium.

749 Page, A.L., and Bingham, F.T. "Cadmium Residues in the Environment." RESIDUE REVIEWS 48 (1973): 1-44.

Excellent review and bibliography. Cadmium hazardous, particularly in area of smelters. Vegetation concentrates cadmium from atmosphere and from root absorption. High residues localized. Authors conclude that more information is needed on cycling of cadmium in the environment.

750 Rowe, Donald R., and Massaro, Edward J. "Cadmium Uptake and Time Dependent Alternations in Tissue Levels in the White Catfish *Ictalurus catus* (Pisces: *Ictaluridae*)." BULLETIN OF ENVIRONMENTAL CONTAMINATION AND TOXICOLOGY 11 (1974): 244-49.

Single dose of 0.2 ppm $CdCl_2$ administered. Liver and kidney concentrate most. Many organs implicated, particularly blood and muscle, but blood levels low.

751 Thomann, Robert S., et al. "A Food Chain Model of Cadmium in Western Lake Erie." WATER RESEARCH 8 (1974): 841-49.

Dwells heavily on modeling properties.

752 Van Hook, R.I. "Cadmium, Lead, and Zinc Distributions between Earthworms and Soils: Potentials for Biological Accumulation." BULLETIN OF ENVIRONMENTAL CONTAMINATION AND TOXICOLOGY 12 (1974): 509-12.

Lead is not accumulated but both cadmium and zinc are, cadmium substantially higher. Fairly predictable rates of concentration in earthworms. Could be used as pollution indicator.

753 Van Hook, R.I., and Yates, A.J. "Transient Behavior of Cadmium in a Grassland Arthropod Food Chain." ENVIRONMENTAL RESEARCH 9 (1975): 76-83.

This is one of few studies on cadmium in ecosystems: authors refer to atmospheric pollution from smelters as large source. Both soluble and insoluble forms of cadmium used. Crickets (herbivorous) reached equilibrium levels in nine days; spiders (predators) did not in thirty. Food chain concentration of cadmium did not occur.

COPPER

754 Fisher, Gerald L. "Function and Homeostasis of Copper and Zinc in Mammals." SCIENCE OF THE TOTAL ENVIRONMENT 4 (1975): 373-412.

Sets a general pattern for the study of trace elements, particularly as levels are altered through contamination. Lists health effects of twelve hazardous trace metals. Most are of industrial origin. Enzymatic actions of metals in blood serum are an important aspect of pollution. Excellent review; has 176 references.

755 McIntosh, A.W. "Fate of Copper in Ponds." PESTICIDES MONITORING JOURNAL 8 (1975): 225-31.

Considering that copper compounds are relatively old in practical use, there are surprisingly few good descriptions of survival and distribution of copper ions. Initially, copper (in copper sulfate) applied to ponds declined to background levels. Then, after reduction, it first concentrated in sediments; subsequently aquatic plants and filamentous algae accumulated very high levels. Later, macrophytes and sunfish picked up copper, but returned to natural levels in two months.

756 Stephenson, R.R., and Taylor, D. "The Influence of EDTA on the Mortality and Burrowing Activity of the Clam (Venerupis decussata) Exposed to Sub Lethal Concentrations of Copper." BULLETIN OF ENVIRONMENTAL CONTAMINATION AND TOXICOLOGY 14 (1975): 304-8.

Heavy metal ions are often acutely damaging to marine organisms. But chronic effects under natural conditions are little known. Copper ion alone is very toxic but toxicity is greatly reduced by binding copper ions to a large organic molecule.

LEAD

757 Benson, W.W., et al. "Lead Poisoning in a Bird of Prey." BULLETIN OF ENVIRONMENTAL CONTAMINATION AND TOXICOLOGY 11 (1974): 105-8.

Prairie falcon (Falco mexicanus) fed heads of waterfowl shot during season died apparently from lead poisoning. Implies lead in brain tissue of waterfowl. Secondary poisoning.

758 Campbell, Irene R., and Mergard, Estelle G. BIOLOGICAL ASPECTS OF LEAD: AN ANNOTATED BIBLIOGRAPHY. 2 vols. Issued by Environmental Protection Agency. Washington, D.C.: 564 p.; 371 p.

Long analyses in annotations. Cross referencing by subject index.

759 Chow, Tsaihwa J., et al. "Lead Pollution: Records in Southern California Coastal Sediments." SCIENCE 181 (1973): 551-52.

Compared with older sediment cores, offshore basins have increased lead levels; due mainly to atmospheric dissemination of combusted additives in gasoline.

760 Ewing, Ben B., and Pearson, John E. "Lead in the Environment." ADVANCES IN ENVIRONMENTAL SCIENCE AND TECHNOLOGY 3 (1974): 1-126.

Major review; nine sections; one hundred and twenty references. Discusses concentrating factors; physiological effects on plants and animals.

761 Hepple, Peter, ed. LEAD IN THE ENVIRONMENT. Essex, Engl.: Applied Science Publishers, 1972. 82 p.

Proceedings of a conference sponsored by the Institute of Petroleum, London, held at the Zoological Society of London, Regent's Park. Seven articles; chiefly on human effects, but also articles on sources of lead, airborne aspects, and significance of plant uptake of lead to animals.

762 Jefferies, D.J., and French, M.C. "Lead Concentrations in Small Mammals Trapped on Roadside Verges and Field Sites." ENVIRONMENTAL POLLUTION 3 (1972): 147-56.

Lead (from gasoline) accumulates in small mammals (Microtus, Clethrionomys, Apodemus) near roadways in amounts ranging from 4.19 to 5.98 ppm (mean values in Great Britain).

763 Khan, M.A.Q., et al. "Lead Content of Soils along Chicago's Eisenhower and Loop-Terminal Expressways." ARCHIVES OF ENVIRONMENTAL CONTAMINATION AND TOXICOLOGY 1 (1973): 209-23.

Lead content strongly correlates with traffic volume and season. Soil as much as 7,600 ppm at forty-five feet from Eisenhower freeway, and 900 ppm at ninety feet. Authors suggest creation of buffer zones in which human beings will not be allowed to work.

764 Mierau, Gary W., and Favara, Blaise E. "Lead Poisoning in Roadside Populations of Deer Mice." ENVIRONMENTAL POLLUTION 8 (1975): 55-64.

Lead concentration in soil and vegetation significantly greater at 15 m from major highway than at 800 m. Bones from mice at 15 m contained ten times as much lead than those at 800 m. Poisoning not demonstrable. Laboratory experiments showed that about five times more lead in tissue than found in wild-taken animals would be required to produce toxic effects. Good references on toxic effects of lead.

765 Patterson, Clair C. "Contaminated and Natural Lead Environments of Man." ARCHIVES OF ENVIRONMENTAL HEALTH 11 (1965): 344-60.

Early review article of amounts and source of lead in the environment. Humans receive about thirty times their "natural" exposure from industrial and transport sources.

766 Welch, W.R., and Dick, D.L. "Lead Concentrations in Tissues of Roadside Mice." ENVIRONMENTAL POLLUTION 8 (1975): 15-21.

Lead in soils related to proximity to highway, but not to traffic volume. Lead in liver, kidney, and bone related to both traffic volume and nearness to road. Lead in brain, lung, stomach, and muscle not noticeably related. No evidence that saturation levels of lead reached.

767 Williamson, P., and Evans, P.R. "Lead: Levels in Roadside Invertebrates and Small Mammals." BULLETIN OF ENVIRONMENTAL CONTAMINATION AND TOXICOLOGY 8 (1972): 280-88.

Authors examined inorganic lead in soils and tissues of many kinds of organisms near English highways. Invertebrates generally below 50 ppm, but woodlice (isopods) nearly 700 ppm. Values in four species of small mammals did not exceed 30 ppm. No evidence of decreasing populations of invertebrates with increasing concentrations of lead, nor of trophic concentration.

MERCURY

768 Annett, C.S., et al. "Mercury in Fish and Waterfowl from Ball Lake, Ontario." JOURNAL OF ENVIRONMENTAL QUALITY 4 (1975): 219-22.

The pollution problem concerns the downstream contamination from a paper mill and chlor-alkali plant, and to the formation of methylmercury and its storage in animals. Average values of mercury in walleye and pike about 5.5 mg/kg. Somewhat higher in livers of waterfowl, but some values extremely high. This is almost a duplication of the conditions at Minamata Bay, Japan.

769 Aulerich, R.J., et al. "Effects of Dietary Mercury on Mink." ARCHIVES OF ENVIRONMENTAL CONTAMINATION AND TOXICOLOGY 2 (1974): 43-51.

Five ppm methylmercury lethal to adults by one month; 10 ppm inorganic (mercuric chloride) had no adverse effects in five months. Clinical symptoms described.

770 Bakir, F., et al. "Methylmercury Poisoning in Iraq." SCIENCE 181 (1973): 230-41.

Massive deaths and illness in Iraq as the result of eating seed grains treated with methylmercury antifungal compounds. Similar incidents in Nigeria and Pakistan.

771 Batti, R., et al. "Methylmercury in River Sediments." CHEMOSPHERE 4 (1975): 13-14.

Sediments were taken from the Seine in France and from three

Italian rivers. Total mercury was high but methylmercury in sediments was surprisingly low; mostly inorganic. But methylmercury was very high in biota. Aquatic organisms must figure in the transformation. All rivers drain heavily industrialized or mining areas.

772 Beckert, W.F., et al. "Formation of Methylmercury in a Terrestrial Environment." NATURE 249 (1974): 674-75.

Mercuric salts, under agricultural conditions, can be transformed into methylmercury.

773 Berg, W., et al. "Mercury Content in Feathers of Swedish Birds for the Last 100 Years." OIKOS 17 (1966): 71-83.

Significant paper. Hawks, owls, eagles, finches, pheasant, partridge examined. Roughly constant mercury levels from 1840 to 1940; differing levels of mercury related to position on food chain; ten- to twentyfold increase in mercury levels in 1940s and 1950s. Chiefly responsible were alkyl-Hg seed dressings, the use of which is now banned.

774 Blair, W., et al. "Application of a Gas Chromatograph-Atomic Adsorption Detection System to a Survey of Mercury Transformations by Chesapeake Bay Microorganisms." CHEMOSPHERE 4 (1974): 167-74.

Ordinarily, bacterial methylation of mercury occurs under anaerobic conditions. But in natural sediments it can also occur in aerobic conditions. Present knowledge of methylation process incomplete.

775 Borg, K., et al. "Experimental Secondary Methylmercury Poisoning in the Goshawk (Accipiter g. gentilis L)." ENVIRONMENTAL POLLUTION 1 (1970): 91-104.

Chickens which had been fed mercury-dressed wheat (8 ppm mercury) were fed to goshawks. Four hawks died within one to one and a half months on such a diet. Clinical and histological manifestations of mercury poisoning described. Residues in tissues as methylmercury reached remarkably high figure in gonads (280 ppm, wet weight, in testes). Fatal level of residues in brain tissue on the order of 30-40 ppm. This study provides direct evidence that methylmercury accumulates in food chains without loss of toxic quality.

776 California, State of. Resources Agency. MERCURY IN CALIFORNIA. Sacramento: State Department of Public Health, 1973. 127 p. Appendices.

Unusually complete coverage for a report of this kind; usable well beyond California. Centers on public health but includes many other aspects. One hundred and twenty-seven references. Mercury occurs naturally in California and is mined. But usage in industry, agriculture, and gold mining contribute to "background" levels. In California these are higher than in most of the United States, averaging between 20 and 40 ppb.

777 Cross, Ronald J., and Jenkins, C. Michael. "Chemical Studies Relating to Environmental Organomercurials." ENVIRONMENTAL POLLUTION 8 (1975): 179-84.

Biological methylation of inorganic mercury is well known. Question here is "what happens further chemically to organomercurials in biological tissue?" Detailed chemical paper.

778 Dieter, Michael P., and Ludke, J. Larry. "Studies on Combined Effects of Organophosphates and Heavy Metals in Birds. I. Plasma and Brain Cholinesterase in Coturnix Quail Fed Methylmercury and Orally Dosed with Parathion." BULLETIN OF ENVIRONMENTAL CONTAMINATION AND TOXICOLOGY 13 (1975): 257-62.

The toxicity of parathion is increased in birds chronically exposed to mercury compounds, and combinations of the two kinds of chemicals become lethal at levels that are by themselves innocuous.

779 D'Itri, Frank M. THE ENVIRONMENTAL MERCURY PROBLEM. Report authorized by H.R. 424, Michigan House of Representatives, Great Lakes Contamination Committee. East Lansing: Michigan State University, 1971. iv, 289 p. (one side only). Mimeo.

Extremely good summation of environmental mercury problem. Nine chapters and 550 references.

780 Eades, J.F. "Pesticides Residues in the Irish Environment." NATURE 201 (1966): 650-52.

Tissues of foxhound, salmon, trout, and several species of birds examined for mercury and chlorinated hydrocarbon (lindane, aldrin, dieldrin, and DDT) insecticides. Pigeons and pheasants had organic mercury residues to which death was attributed. Most other tissues had organochlorine residues. In eggs of five bird species, two had mercury and all had organochlorine residues.

781 Edwards, T., and McBride, B.C. "Biosynthesis and Degradation of Methylmercury in Human Feces." NATURE 253 (1975): 462-64.

Microbial flora in the intestine of man has the potential to transform Hg^{2+} to highly toxic methylmercury. This could add to body burden, increasing the severity of poisoning.

782 Elder, John A., and Gaufin, Arden R. "The Toxicity of Three Mercurials to Pteronarcys californica Newport, and Some Possible Physiological Effects Which Influence the Toxicities." ENVIRONMENTAL RESEARCH 7 (1974): 169-75.

Toxicities of phenylmercuric, methylmercuric, and mercuric chloride to stoneflies. Decreasing toxicity is in order named. Need to know most susceptible stages in aquatic organisms.

783 Fagerstroem, T[orbjoern]., and Aesell, B. "Methyl Mercury Accumulation in an Aquatic Food Chain. A Model and Some Implications for Research Planning." AMBIO 2 (1973): 164-71.

Cuts in industrial mercury effluents in Sweden should have led to reduced mercury in aquatic organisms. But buildup is increasing in methylated form. Paper has mathematic model of three-step food chain. Authors suggest that direct uptake of methylmercury from water needs study.

784 Fagerstroem, T[orbjoern]., and Jerneloev, A. "Biological Methylation of Mercury, Food Chain Accumulation, Ecological Effects, and Routes of Exposure to Man." CRITICAL REVIEWS IN ENVIRONMENTAL CONTROL 5 (1974): 296-305.

A generalized but useful review, with graphic models illustrating ecosystem flow and chemical changes.

785 Fang, S.C. "Uptake and Biotransformation of Phenylmercuric Acetate by Aquatic Organisms." ARCHIVES OF ENVIRONMENTAL CONTAMINATION AND TOXICOLOGY 1 (1973): 18-26.

Guppy, snail, elodea, and coontail rapidly pick up ^{203}Hg-labeled PMA (phenylmercuric acetate) from water. Not quickly eliminated. Biological half-life is forty-three to fifty-eight days.

786 Fimreite, N[orvald]. "Mercury Contamination of Aquatic Birds in Northwestern Ontario." JOURNAL OF WILDLIFE MANAGEMENT 38 (1974): 120-31.

Contaminant source is a chlorine plant, some fifty to eighty miles from collection sites. Values in eggs and tissues are on order of 100 ppm in some scavenger and fish-eating birds. Ratio of total mercury to methylmercury lower than in most reports. Breeding of loons seemingly severely restricted but breeding of terns is not. Major article with detailed tables and good bibliography.

787 Fimreite, N[orvald]., and Karstad, L. "Effects of Dietary Methyl Mercury on Red-tailed Hawks." JOURNAL OF WILDLIFE MANAGEMENT 35 (1971): 293-300.

Six groups of three birds each were fed chicks contaminated with mercury. After one month's feeding there was mortality in hawks (Buteo jamaicensis) on most contaminated diet. Mercury levels in livers of hawks that died were about 20 ppm. Many neurological symptoms were also exhibited in birds that died or showed serious external pathology.

788 Fimreite, N., and Reynolds, L.M. "Mercury Contamination of Fish in Northwestern Ontario." JOURNAL OF WILDLIFE MANAGEMENT 37 (1973): 62-68.

Study of several species of fishes upstream and downstream from a chlorine plant. Maximum value, 27.8 ppm, in pike taken 50-60 miles downstream. Elevated levels as much as 200 miles downstream. Background levels upstream and in oligotrophic (sterile) lakes are rarely as much as 1.0 ppm.

789 Fimreite, N., et al. "Mercury Contamination of Canadian Prairie Seed Eaters and Their Avian Predators." CANADIAN FIELD-NATURALIST 84 (1970): 269-76.

Mercury is common in livers of seed-eating birds and their prey. Low in most mammals; high in bird-eating hawks. Eggs reflect adult tissue levels. Authors suggest that mercury contributes to eggshell thinning, along with chlorinated hydrocarbons.

790 ______. "Mercury in Fish and Fish-eating Birds Near Sites of Industrial Contamination in Canada." CANADIAN FIELD-NATURALIST 85 (1971): 211-20.

Near a chlor-alkali plant. Mercury levels exceeding 0.5 ppm (legal tolerance) in almost all pike. Values in pike, pumpkin seed, and lake trout ranged to 10.5 ppm. Body weight in fishes corresponds to mercury level; so also does trophic position in birds. Highest mercury level was 17.4 ppm in red-necked grebe.

791 Friberg, Lars, and Vostal, Jaroslav, eds. MERCURY IN THE ENVIRONMENT: AN EPIDEMIOLOGICAL AND TOXICOLOGICAL STUDY. Cleveland, Ohio: CRC Press, 1972. 215 p.

Ten chapters; six contributors. General coverage and technical reviews. Excellent reference.

792 Hartung, Rolf, and Dinman, Bertram D., eds. ENVIRONMENTAL

MERCURY CONTAMINATION. Ann Arbor, Mich.: Ann Arbor Science Publishers, 1972. ix, 349 p.

Many contributors; five parts. International coverage on occurence, methods of analysis, environmental dynamics, and biological effects.

793 Heinz, Gary H. "Effects of Low Dietary Levels of Methyl Mercury on Mallard Reproduction." BULLETIN OF ENVIRONMENTAL CONTAMINATION AND TOXICOLOGY 11 (1974): 386-92.

Mallards on a diet of 0.5 or 3 ppm methylmercury dicyandiamide. Affected reproduction at 3 ppm but did not affect adults during twelve months of dosage. Reproductive effects included duckling mortality, reduced egg laying, and increased embryonic mortality. Control eggs heavier than those at either mercury level in diet but no eggshell thinning associated with mercury. Residues in eggs reached 1 ppm (from 0.5 ppm diet) and 6-9 ppm (from 3 ppm diet).

794 Hollins, J.G., et al. "The Whole Body Retention and Tissue Distribution of [^{203}Hg] Methylmercury in Adult Cats." TOXICOLOGY AND APPLIED PHARMACOLOGY 33 (1975): 438-49.

This study relates experimentally to differential storage and elimination of methylmercury poisoning in man; prompted by several outbreaks of such poisoning. Study is descriptive, detailed, and comprehensive. After 156 days (following single oral treatment), over 35 percent of mercury was still in body, mostly in hair and liver.

795 Holm, Harvey W., and Cox, Marilyn F. MERCURY IN AQUATIC SYSTEMS: METHYLATION, OXIDATION-REDUCTION, AND BIO-ACCUMULATION. Ecological Research Series, National Environmental Research Center, Corvallis, Oregon. Washington, D.C.: Government Printing Office for Environmental Protection Agency, 1974. 38 p.

Evaluation of the fate of mercury in aquatic environments, with emphasis on biota and on transfer of inorganic to organic mercury. Good literature.

796 Huckabee, John W., and Janzen, S.A. "Mercury in Moss: Derived from the Atmosphere or from the Substrate." CHEMOSPHERE 4 (1975): 55-60.

Mosses are good pollution indicators because they absorb chemical elements directly from the atmosphere. But do they absorb and translocate elements from their substrate? Answer seems to be that mosses do in fact get most, if not all, mercury from air--as has been assumed all along.

797 Huckabee, J[ohn]. W., et al. "Distribution of Mercury in Pheasant Muscle and Feathers." JOURNAL OF WILDLIFE MANAGEMENT 36 (1972): 1306-9.

Seventy-five percent of wild birds had detectable mercury. Thirty-five percent had values exceeding 0.5 ppm (U.S. legal limit). Mercury in feathers loosely correlated with levels in other tissue.

798 Iverson, F., and Hierlihy, S.L. "Biotransformation of Methyl Mercury in Guinea Pig. BULLETIN OF ENVIRONMENTAL CONTAMINATION AND TOXICOLOGY 11 (1974): 85-91.

Amount of biotransformation inversely proportional to dose rate. Kidney binds bulk of inorganic mercury. Lower levels of inorganic mercury in liver.

799 Jerneoloev, A. "Conversion of Mercury Compounds." In CHEMICAL FALLOUT, pp. 68-74. Springfield, Ill.: Charles C. Thomas, Publisher, 1969.

Several types of mercury--metallic or organic--can be converted to methylmercury in fish.

800 _____. "Mercury--A Case Study of Marine Pollution." In THE CHANGING CHEMISTRY OF THE OCEANS, edited by David Dyrssen and Daniel Jagner, pp 161-69. New York: John Wiley & Sons, 1972.

Factors explaining rapid methylation of mercury from simpler organic and inorganic sources, whether natural or man-caused.

801 Johnels, A.G., and Westermark, T. "Mercury Contamination of the Environment in Sweden." In CHEMICAL FALLOUT, edited by Morton W. Miller and George G. Berg, pp. 221-41. Springfield, Ill.: Charles C. Thomas, Publisher, 1969.

Study of mercury content of feathers in eleven terrestrial bird species. Traced back some 100 years in museum specimens. There was ten to twenty times more mercury after 1940.

802 Johnels, A. G., et al. "Pike (*Esox lucius* L.) and Some Other Aquatic Organisms in Sweden as Indicators of Mercury Contamination in the Environment." OIKOS 18 (1967): 323-33.

Mercury in pike is concentrated 3,000 or more times the level in water. Human activities raise mercury levels much above natural.

803 Kamps, LaVerne R., et al. "Total Mercury--Monomethylmercury Content of Several Species of Fish." BULLETIN OF ENVIRONMENTAL CONTAMINATION AND TOXICOLOGY 8 (1972): 273-79.

In edible portions of five species of fish mercury content ranged from 0.04 to 2.6 ppm, chiefly as monomethylmercury.

804 Katz, Arthur. "Mercury Pollution: The Making of an Environmental Crisis." CRITICAL REVIEWS IN ENVIRONMENTAL CONTROL 2 (1972): 517-35.

General survey of effects and the socioadministrative problems connected.

805 Kim, Ke Chung, et al. "Mercury in Tissues and Lice of Northern Fur Seals." BULLETIN OF ENVIRONMENTAL CONTAMINATION AND TOXICOLOGY 11 (1974): 281-83.

Article documents the first detection of mercury in nursing cows, newborn and suckling pups, and in two species of sucking lice of Callorhinus ursinus. Mercury in all samples. Hair contained highest levels, with 5.36 ppm in two-month old pups. Lice from these pups at 0.63 ppm mercury. Particularly useful literature list.

806 Knauer, George A., and Martin, John H. "Mercury in a Marine Pelagic Food Chain." LIMNOLOGY AND OCEANOGRAPHY 17 (1972): 868-76.

Phyto- and zooplankton and anchovies from Monterey Bay, California, showed low levels of mercury and no indication of food chain amplification over a ten-month period.

807 Knight, Luther A., Jr., and Harvey, Edward D., Sr. "Mercury Residues in the Common Pigeon (Columba livia) from the Jackson, Mississippi, Area--1972." PESTICIDES MONITORING JOURNAL 8 (1974): 102-4.

Median concentration of total mercury in pigeon brain was 22 ppb and in claws, 14-85 ppm. Sources possible are treated grains, contaminated weed seeds, and naturally occurring mercury.

808 Kolb, L.P., et al. ECOLOGICAL IMPLICATIONS OF DIMETHYL MERCURY IN AN AQUATIC FOOD CHAIN. Logan: Utah State University, 1973. 50 p. Figures and tables.

Useful summary. Concentration of dimethyl mercury in biota 30-90 times greater than in water. But concentration is 100 times more with monomethyl form. Dimethyl form seemingly innocuous, but monomethyl dangerous.

809 Kopfler, Frederick C. "The Accumulation of Organic and Inorganic Mercury Compounds by the Eastern Oyster (Crassostrea virginica)."

BULLETIN OF ENVIRONMENTAL CONTAMINATION AND TOXICOLOGY 11 (1974): 275-80.

Continuous exposure of even 1.0 microg/l mercury to adult oysters resulted in rapid concentrations exceeding 0.5 ppm, the "action guideline" established by the Food and Drug Administration.

810 Krenkel, P.A. "Mercury: Environmental Considerations." Parts 1 and 2. CRITICAL REVIEWS IN ENVIRONMENTAL CONTROL 3 (1974): 303-73; 4 (1974): 251-339.

Comprehensive review of chemistry, environmental distribution, and briefly, biological and ecological effects. Long reference list.

811 Martin, William E. "Mercury and Lead Residues in Starlings--1970." PESTICIDES MONITORING JOURNAL 6 (1972): 27-32.

Starlings have been used since 1967 by the U.S. Fish and Wildlife Service as an indicator species, monitored regularly for environmental pollutants. In this study mercury residues were all below 0.5 ppm; lead residues ranged from 0.4 to 13.3 ppm, with mean of 3.18 ppm.

812 Miller, G.E., et al. "Mercury Concentrations in Museum Specimens of Tuna and Swordfish." SCIENCE 175 (1972): 1121-22.

Mercury levels of museum specimens of seven tuna caught sixty-two and ninety-three years ago, and a swordfish caught twenty-five years ago showed levels of mercury in the same range as recently caught specimens. Mean mercury concentration, 3.1 ppm, dry weight. This concentration shows contrast to freshwater environments, in which mercury levels have risen.

813 Mottet, N. Karle. "Effects of Chronic Low-Dose Exposure of Rat Fetuses to Methylmercury Hydroxide." TERATOLOGY 10 (1974): 173-90.

Dose of below 15 microg/g to near-term pregnant rats had no effects on fetuses. But doses up to 75 microg/g produced a dose-response relationship resulting in maternal death. Fetal resorption, and decreasing fetal weight. No gross, microscopic, or behavioral changes were observable.

814 Nelson, J.D., Jr., and Colwell, R.R. "The Ecology of Mercury-Resistant Bacteria in Chesapeake Bay." MICROBIAL ECOLOGY 1 (1975): 191-218.

Major article in a new journal. Many references. Plate

counts of methyl-resistant bacteria are valid index of mercuric ions in natural systems.

815 Neville, G.A., and Berlin, M. "Identification and Biotransformation of Organomercurial Compounds in Living Systems." ENVIRONMENTAL RESEARCH 7 (1974): 75-82.

Brief review; good reference list. Toxic effects in environment are varied; must know more about identification and isolation of compounds.

816 Olson, Gayle F., et al. "Mercury Residues in Fathead Minnows, Pimephales promelas Rafinesque, Chronically Exposed to Methylmercury in Water." BULLETIN OF ENVIRONMENTAL CONTAMINATION AND TOXICOLOGY 14 (1975): 129-34.

Fish take up methylmercury from water and concentrate residues in tissues up to fortyfold, differentially settling in muscle. Hence, it is possible to exceed 0.5 ppm legal tolerance in edible parts even when organic mercury is in water in small amounts.

817 Olson, Kirsten, and Boush, G. Mallory. "Decreased Learning Capacity in Rats Exposed Prenatally and Postnatally to Low Doses of Mercury." BULLETIN OF ENVIRONMENTAL CONTAMINATION AND TOXICOLOGY 13 (1975): 73-79.

Methymercury easily crosses the placental barrier and accumulates in fetuses. The most damaging feature of mercury toxicity comes from this fact. The offspring of rats fed 2 ppm mercury in diet showed deleterious effects on behavioral maturation and learning ability. Authors call for more research on effects of naturally occurring mercury and of the toxicity of mercury in combination with other metals.

818 Peakall, David B., and Lincer, Jeffrey L. "Methyl Mercury: Its Effects on Eggshell Thickness." BULLETIN OF ENVIRONMENTAL CONTAMINATION AND TOXICOLOGY 8 (1972): 89-90.

Methylmercury in diet (to 10 ppm) did not produce eggshell thinning in kestrels and ring doves. It does in Coturnix quail, when given as mercuric chloride up to 8 ppm in diet.

819 Reimer, Amada A., and Reimer, Roger D. "Total Mercury in Some Fish and Shellfish along the Mexican Coast." BULLETIN OF ENVIRONMENTAL CONTAMINATION AND TOXICOLOGY 14 (1975): 105-11.

In Canada, Japan, Sweden, and the United States, fish samples frequently exceed the 0.5 ppm tolerance limit. On both coasts of Mexico, in fish, crustaceans, and molluscs, no seafood sold commercially had as much as 0.5 ppm. Most contain mercury at about 0.05 ppm, although even at that level, accumulation occurs in tissues.

820 Rivers, Jerry B., et al. "Total and Organic Mercury in Marine Fish." BULLETIN OF ENVIRONMENTAL CONTAMINATION AND TOXICOLOGY 8 (1972): 257-66.

Nine species tested; blue marlin highest. Important to consider naturally occurring mercury and balance between total and methylated forms. Authors suggest that demethylation occurs in marlin when total mercury reaches certain level.

821 Roales, R.R., and Perlmutter, A. "Toxicity of Methyl Mercury and Copper Applied Singly and Jointly, to the Blue Gourami, Trichogaster trichopterus." BULLETIN OF ENVIRONMENTAL TOXICOLOGY 12 (1974): 633-39.

In these fish 96-hour-TL_m: methylmercury, 89.5 ppb; copper, 91.2 ppb. But together, copper seems partially antagonistic to methylmercury poisoning.

822 Saha, J.G. "Significance of Mercury in the Environment." RESIDUE REVIEWS 42 (1972): 104-63.

Significant and comprehensive article. Long bibliography. Stimulated by Canadian concerns, but international in scope. Generally (in the United States, Canada, Japan, and Sweden) "allowable daily intake" of 0.7 mg/week is much below tolerance limits in foods of 0.5 or 1.0 ppm.

823 Sell, Jerry L., et al. "Concentration of Mercury in Animal Products and Soils in North Dakota." ARCHIVES OF ENVIRONMENTAL CONTAMINATION AND TOXICOLOGY 3 (1975): 278-88.

Detailed survey in humans and in animal foods relating to agricultural practices. Most values low; strangely, the highest was in head hair of women (the only value to exceed 1.0 ppm).

824 Spangler, William J., et al. "Methylmercury: Bacterial Degradation in Lake Sediments." SCIENCE 180 (1973): 192-93.

Heat in lake sediments increases amount of methylmercury to a point, then amount decreases as mercury volatizes. Bacteria can reduce organic mercury to methane and Hg.

825 Spann, J.W., et al. "Ethyl Mercury P-Toluene Sulfonanilide: Lethal and Reproductive Effects on Pheasants." SCIENCE 175 (1972): 328-30.

Chemical (in Ceresan 57) at 30 ppm in diet is lethal to ring-necked pheasants. Egg production and survival of third-week embryoes are sharply reduced at 10 ppm in diet.

826 Sumner, Arthur K., et al. "Mercury Residues in Fish from Saskatchewan Waters with and without Known Sources of Pollution--1970." PESTICIDES MONITORING JOURNAL 6 (1972): 122-25.

Levels of mercury were higher in fish (0.18-8.88 ppm) caught downstream from a chlor-alkali plant than those (0.25-1.2 ppm) caught upstream where pollution sources are nonindustrial.

827 Tejning, S. "Mercury in Pheasants (Phasianus colchicus L.) Deriving from Seed Grain Dressed with Methyl and Ethyl Mercury Compounds." OIKOS 18 (1967): 334-44.

Remarkably high mercury levels in pheasants and other seed-eating birds explained by seeds sowed with about 16 mg/kg mercury as fungicide. Shows up in more concentrated form in predators. This article was one of the original announcements of the "mercury problem" in Sweden.

828 U.S. Department of Health, Education, and Welfare. Special Report to the Secretary's Pesticide Advisory Committee. "Hazards of Mercury." ENVIRONMENTAL RESEARCH 4 (1971): entire issue.

Presented as one long well-written paper with many references and appendixes. Excellent summation to 1970.

829 Vermeer, K[ees]., and Armstrong, F.A.J. "Mercury in Canadian Prairie Ducks." JOURNAL OF WILDLIFE MANAGEMENT 36 (1972): 179-82.

Residue levels over 0.5 ppm were found only in common mergansers. Livers on average had 2.9 times the mercury found in breast muscle. Primary feathers had 22 times as much in adult pintail ducks than in immatures. Mercury source at that time must have been mercury-treated seed grain.

830 Vermeer, K[ees]., et al. "Mercury in Aquatic Birds at Clay Lake, Western Ontario." JOURNAL OF WILDLIFE MANAGEMENT 37 (1973): 58-61.

Total mercury levels ranging from 2-16 ppm in eggs did not affect hatching and fledging in herring gulls. Majority of mercury methylated in several species of ducks. Crayfish muscle had highest mercury levels among common food items.

831 Westermark, T., et al. "Mercury Content of Bird Feathers before and after Swedish Ban on Alkyl Mercury in Agriculture." AMBIO 4 (1975): 87-92.

Mercury level in feathers is determined when feather is formed; fixed thereafter. Strong indications of a decrease. But mercury type changed, and there is some resumption of use of seed dressings. Accumulated quickly by seed-eating birds and passed on to predators. Stored as methylmercury.

832 Westoo, Gunnell. "Methylmercury as Percentage of Total Mercury in Flesh and Viscera of Salmon and Sea Trout of Various Ages." SCIENCE 181 (1973): 567-68.

Proportion of methylmercury to total mercury averaged 93 percent in fish of all ages checked. Range was 81-98 percent. In viscera the range was 26-27 percent, without age dependence.

833 Wood, J.M., et al. "Synthesis of Methylmercury Compounds on Extracts of a Methanogenic Bacterium." NATURE 220 (1968): 173-74.

Establishes that anaerobic bacteria are responsible for conversion of organic mercury compounds from inorganic; mercury accumulates in living tissues in organic form.

SELENIUM

834 Gauther, H.E., et al. "Selenium: Relation to Decreased Toxicity of Methylmercury Added to Diets Containing Tuna." SCIENCE 175 (1972): 1122-24.

Japanese quail given 20 ppm methylmercury diets containing 17 percent tuna survived longer than those given this concentration in corn-soya diet. Tuna has relatively high content of selenium and tends to add selenium when mercury is present. Selenium content comparable to that in tuna tended to decrease methylmercury toxicity in rats. Selenium in tuna may lessen danger to man of mercury in tuna.

835 Gissel-Nielson, G., and Gissel-Nielsen, M. "Ecological Effects of Selenium Application to Field Crops." AMBIO 2 (1973): 114-17.

Selenium is an elemental nutrient, but at higher concentrations is toxic. In selenium-deficient areas, addition of selenite was considered. Biological half-life (days) for some animals shown by the decay rate of the short-lived radioisotope ^{75}Se. Mean values of selenium in animals experimentally tested: bank vole--13; pheasant--10; eel--27; guppy--13; ant--15; leech--28; earthworm--64. Conclude no hazard; selenium is actually a nutrient replacement in high-yield crops.

836 Ohi, Gen, et al. "Interaction of Dietary Methylmercury and Selenium on Accumulation and Retention of These Substances in Rat Organs." TOXICOLOGY AND APPLIED PHARMACOLOGY 32 (1975): 527-33.

Concentrations in animal tissues of both methylmercury and selenium unusually high in vicinity of Minamata, Japan (where mercury poisoning had occurred on a large scale). This work shows that selenium does aid in methylmercury retention; that inorganic selenium is not retained long but organic forms can be; and that "selenite" does, to noticeable degree, protect against the toxicity of organic mercury. A good example of chemical interaction (explanation not known) in two well-known toxic materials.

837 Pakkala, Irene S., et al. "A Survey of the Selenium Content of Fish from 49 New York State Waters." PESTICIDES MONITORING JOURNAL 6 (1972): 107-14.

Concentration of selenium from many species usually below 1 ppm. No correlation of selenium level and age and sex of fish, nor did it seem to be accumulative in lake trout.

838 Stadtman, Thressa C. "Selenium Biochemistry." SCIENCE 183 (1974): 915-22.

Toxic effects of selenium have been known for a long time. Only in past few years has it been recognized as an essential micronutrient in animals and bacteria. Long reference list.

839 Stoeswand, G.S., et al. "Dietary Selenium Protection of Methylmercury Intoxication of Japanese Quail." BULLETIN OF ENVIRONMENTAL CONTAMINATION AND TOXICOLOGY 11 (1974): 152-56.

Sodium selenite at 5 ppm in diet protected against mercury intoxication (totally, when fed together). High mortality when selenium introduction interrupted.

ZINC

840 Bengtsson, Bengt-Erik. "The Effects of Zinc on the Mortality and Reproduction of the Minnow *Phoxinus phoxinus* L." ARCHIVES OF ENVIRONMENTAL CONTAMINATION AND TOXICOLOGY 2 (1974): 342-55.

Fry showed increasing mortality beginning at 0.08 ppm, one-fortieth of the 96-hr LC50 estimated for adults. Long bibliography.

841 Gasaway, William C., and Buss, Irven O. "Zinc Toxicity in the

Mallard Duck." JOURNAL OF WILDLIFE MANAGEMENT 36 (1972): 1107-17.

Zinc toxicity in waterfowl due to accumulations in rivers downstream from smelting plant effluents. Mallards fed 3,000-12,000 ppm zinc showed drastic effects on reproductive organs, adrenals, kidneys, and body weight. Waterfowl die-offs related to zinc toxicosis. Detailed article with good bibliography.

842 Holmes, Charles W., et al. "Migration and Redistribution of Zinc and Cadmium in Marine Estuarine System." ENVIRONMENTAL SCIENCE AND TECHNOLOGY 8 (1974): 255-59.

Unusually high levels of zinc and cadmium in Corpus Christi Bay and adjoining harbor. Summer stagnation precipitates metals in harbor; increased water exchange in winter washes them into bay.

843 Nehring, R. Barry, and Goettl, John P., Jr. "Acute Toxicity of a Zinc-polluted Stream to Four Species of Salmonids." BULLETIN OF ENVIRONMENTAL CONTAMINATION AND TOXICOLOGY 12 (1974): 464-69.

Rainbow trout least tolerant. TL50 for four species: rainbow trout, 0.41 mg/l; brown trout, 0.64; cutthroat, 0.67; brook trout, 0.96.

Section 11

SYNTHETIC ORGANIC COMPOUNDS

We view synthetic compounds as essential to the manufacture of products required of technological society. They are intentionally synthesized, manufactured, and distributed. But, as waste products they become unintentional pollutants, universally contaminating, leading to ends we cannot foresee. As the cited articles indicate, these products are generally toxic and they are relatively stable. Many of them are chlorinated hydrocarbons as previously described under pesticides. Because of their chemical similarities, many of their unknown effects in the environment can be conjectured to be much like those of DDT.

But two aspects of their ubiquitous environmental presence stand out. The first is that, along with other synthetic chlorinated hydrocarbons, these chemicals are novel in earth's evolutionary history because their degradation and removal depend upon organisms that have no selectively derived basis for so doing; and these organisms must confront the toxicity of these chemicals as well. Time is required for those evolutionary adjustments, and these compounds have become major contaminants only in the last two or three decades. The second remarkable aspect of synthetic compounds is that their volume has become immense in such a short time. Billions of pounds of these products related to the plastics industry have been manufactured and distributed in an historically very short space of time. No one can say what the total amounts distributed throughout the earth's biosphere are. Nor can we yet conclude that they are hazardous on a large scale. They are novel, abundant, and universally distributed. They constitute an unwanted global experiment. The end to the experiment is not in sight.

The best known of this class of compounds is the polychlorinated biphenyls (PCBs). Related to them are the polybrominated biphenyls (PBBs) and the polychlorinated terphenyls (PCTs). Among others are the abundant phthalate acids and esters, vinyl chloride, and polyvinyl chloride (PVC). The PCBs are given the most attention in this review.

Synthetic Organic Compounds

GENERAL AND MULTIPLE REFERENCES

844 Autian, John. "Toxicity and Health Threats of Phthalate Esters: Review of the Literature." ENVIRONMENTAL HEALTH PERSPECTIVES, no. 4 (1973), pp. 3-26.

Fine review of plasticizers, identified as environmental pollutants only since 1971. Yet almost a billion pounds are manufactured per year in the United States. Known to concentrate in fish and plankton and to affect growth. Also shows teratogenic and mutagenic effects in mice.

845 Belisle, Andre A., et al. "Analysis of Tissues of Mallard Ducks Fed Two Phthalate Esters." BULLETIN OF ENVIRONMENTAL CONTAMINATION AND TOXICOLOGY 13 (1975): 129-32.

Phthalic ester plasticizers are used primarily in polyvinyl chloride plastics. Related chemicals are used as insecticides and repellents. Production was one billion pounds in the United States in 1972. Although widely distributed contaminants, low levels apparently do not accumulate in mallard ducks.

846 Carpenter, Edward J., and Smith, K.L., Jr. "Plastics on the Sargasso Sea Surface." SCIENCE 175 (1972): 1240-41.

Plastic particles abundant and widely spread in the western Sargasso Sea. Brittle because of weathering. Diatoms and hydroids attach. Increasing production of plastics will produce more such marine waste. Possibly a source of some PCBs observed in marine organisms.

847 Environmental Protection Agency. PRELIMINARY ASSESSMENT OF THE ENVIRONMENTAL PROBLEMS ASSOCIATED WITH VINYL CHLORIDE AND POLYVINYL CHLORIDES (APPENDICES). 2 vols. Report on the Activities and Findings of the Vinyl Chloride Task Force. Washington, D.C.: 1974. 29 p.; 67 p.

Reports on economics, production, regional distribution, analytical methods, and occupational health related to vinyl chloride. Ecological effects, routes of transfer, and bioaccumulation rates unknown.

848 Haley, Thomas J. "Vinyl Chloride: How Many Unknown Problems?" JOURNAL OF TOXICOLOGY AND ENVIRONMENTAL HEALTH 1 (1975): 47-73.

Major review and bibliography of the chemistry, biotransformation, toxicology, and carcinogenicity of vinyl chloride. Hazards largely unknown; calls for intensive research and regulation.

849 Metcalf, Robert L., et al. "Uptake and Fate of Di-2-Ethylhexyl Phthalate in Aquatic Organisms and in a Model Ecosystem." ENVIRONMENTAL HEALTH PERSPECTIVES, no. 4 (1973), pp. 27-34.

DEHP is a micropollutant rapidly biomagnified by a variety of plants and animals. Environmental kinetics resemble DDT.

850 Peakall, David B. "Effects of Di-N-Butyl and Di-2-Ethylhexyl Phthalate on the Eggs of Ring Doves." BULLETIN OF ENVIRONMENTAL CONTAMINATION AND TOXICOLOGY 12 (1974): 698-702.

Esters as commonly used plasticizers found in environment. Reports effects of 10 ppm in diet of each ester on eggshell thickness, breaking strength, permeability, and shell structure of *Streptopelia*. Most significant results: shell thickness decreased and permeability to water increased with first chemical. Thickness returns to normal quickly when chemical withdrawn, in contrast to organochlorines.

851 Rosenberg, Rutger. "Effects of Chlorinated Aliphatic Hydrocarbons on Larval and Juvenile *Balanus balanoides* L." ENVIRONMENTAL POLLUTION 3 (1972): 313-18.

By-products of vinyl chloride production affect barnacle nauplii (larvae) at sublethal levels (2-4 ppm).

852 Scott, T.S. CARCINOGENIC AND CHRONIC TOXIC HAZARDS OF AROMATIC AMINES. Amsterdam and New York: Elsevier Publishing Co., 1962. ix, 209 p.

Chemical group does not occur naturally at all but many compounds used in manufacturing. All contain NH_2 group; aniline, best known. [Only the industrial hazard of these chemicals has thus far been studied, but one wonders if they might become major environmental contaminants.]

853 Sugawara, N. "Effect of Phthalate Esters on Shrimp." BULLETIN OF ENVIRONMENTAL CONTAMINATION AND TOXICOLOGY 12 (1974): 421-24.

Increasing plastics pollution. Checked toxicity of three esters against *Artemia* (brine shrimp) eggs and nauplii. Toxic order (DBP-DEP-DMP) inversely proportional to solubility in water. At levels already found in environment, all three esters are toxic to shrimp eggs and larvae.

854 U.S. Department of Health, Education, and Welfare. "Perspective on PAEs." ENVIRONMENTAL HEALTH PERSPECTIVES, Experimental Issue no. 3, (1973): entire issue (182 p.).

Twenty-five papers reviewing the importance of phthalic acid esters as environmental contaminants.

855 Zitko, V. "Uptake of Chlorinated Paraffins and PCB from Suspended Solids and Food by Juvenile Atlantic Salmon." BULLETIN OF ENVIRONMENTAL CONTAMINATION AND TOXICOLOGY 12 (1974): 406-12.

Chlorinated paraffins are suggested as replacements for PCBs. They are much less accumulated than PCB, if at all. Possibly toxic to juvenile salmon.

POLYCHLORINATED BIPHENYL (PCB)

856 Ahnoff, M., and Josefsson, B. "Polychlorinated Biphenyl (PCB) in Gota River Water." AMBIO 4 (1975): 172-74.

Little is known about waters with low levels PCB. Describes new analytical method when PCB is present only in trace amounts. In this instance it was shown that river played little role in dispensing PCB.

857 Allen, J.R., and Abrahamson, L.J. "Morphological and Biochemical Changes in the Liver of Rats Fed Polychlorinated Biphenyls." ARCHIVES OF ENVIRONMENTAL CONTAMINATION AND TOXICOLOGY 1 (1973): 265-80.

Following administration in diet for six weeks of 0.1 percent PCB in three Arochlor types, liver became enlarged and liver microsomal enzymes increased activity. Three-fold enlargement of liver in animals killed after six weeks on above diet.

858 Ayres, David C., and Denton, Gary R.W. "On the Diversity of Products Obtained during Synthesis of Polychlorbiphenyls by the Van Roosmalen Procedure." BULLETIN OF ENVIRONMENTAL CONTAMINATION AND TOXICOLOGY 14 (1975): 361-69.

Many peaks on chromatograms, as many as fourteen. Illustrates compound nature of PCB and difficulty in analysis.

859 Babish, John G., et al. "Polybrominated Biphenyls: Tissue Distribution and Effect on Hepatic Microsomal Enzymes in Japanese Quail." JOURNAL OF AGRICULTURAL AND FOOD CHEMISTRY 23 (1975): 879-82.

Maximum enzyme induction occurred at 20 ppm intake in males, at 100 ppm in females. Potential toxicity of PBBs as great as PCBs.

860 Bagley, G.E., et al. "Identification of Polychlorinated Biphenyls in Two Bald Eagles by Combined Gas-Liquid Chromatography--Mass Spectrometry." JOURNAL OF THE ASSOCIATION OF OFFICIAL ANALYTICAL CHEMISTS 53 (1970): 251-61.

PCB found in eagle tissues; article concerns technique of separating "interfering" chlorinated hydrocarbons for individual chemical determinations.

861 Bush, Brian, et al. "Toxicity and Persistence of PCB Homologs and Isomers in the Avian System." ARCHIVES OF ENVIRONMENTAL CONTAMINATION AND TOXICOLOGY 2 (1974): 195-212.

Detailed comprehensive experimental review of PCBs in birds. Good bibliography.

862 Cecil, Helene C., et al. "Effects of Polychlorinated Biphenyls and Terphenyls and Polybrominated Biphenyls on Pentobarbitol Sleeping Times of Japanese Quail." ARCHIVES OF ENVIRONMENTAL CONTAMINATION AND TOXICOLOGY 3 (1975): 183-92.

DDT and methoxychlor lengthen sleeping time, but it is shortened with the several Arochlors used in this study. Presumably means increased metabolism of pentobarbitol by induced liver microsomal enzymes. Degree of expression correlated with chlorine content.

863 Clausen, J., et al. "The Content of Polychlorinated Hydrocarbons in Arctic Mammals." BULLETIN OF ENVIRONMENTAL CONTAMINATION AND TOXICOLOGY 12 (1974): 529-34.

Mammals studied were four seal species, polar bear, polar fox, and sheep. DDE and PCB were found in all. Highest level was DDE (1.25 ppm, lipid) in bear which also contained 21.0 ppm PCB; highest PCB in porpoise was 11.4 ppm. Aldrin was present in most samples. Sources are probably algae and fish which concentrate the compounds and which are distributed northward by the Gulf Stream.

864 Craigie, J.S., and Hutzinger, O[tto]. "Effects of Commercial Chlorinated Hydrocarbons and Specific Chlorobiphenyls on the Growth of Seven Species of Marine Phytoplankton." CHEMOSPHERE 3 (1975): 139-44.

Many PCB formulations used to determine effects on growth. Some PCB had no effect; some species depressed, others not depressed at lower ranges. But no species grew at 100 ppm of any PCB. Highly chlorinated biphenyls seem less toxic.

865 Crosby, Donald G., and Moilanen, Kenneth W. "Photodecomposition of Chlorinated Biphenyls and Dibenzofurans." BULLETIN OF ENVIRONMENTAL CONTAMINATION AND TOXICOLOGY 10 (1973): 372-77.

DDT and DDE are converted in part to di-, tri-, and tetrachlorobiphenyls under simulated atmospheric irradiation.

Would be expected to precipitate in rainfall if formed in atmosphere. Possibly degradative effect from ultraviolet component of sunlight. For a review of this work, see T.H. Maugh II in section 5, DDT and Its Derived Relatives, DDD and DDE.

866 Crump-Wiesner, Hans J., et al. "A Study of the Distribution of Polychlorinated Biphenyls in the Aquatic Environment." PESTICIDES MONITORING JOURNAL 8 (1974): 157-61.

Residues ubiquitous in aquatic environment. Range of sampling in nineteen states showed 0.1 microg/kg in water, and from 0.5 microg/kg to 3200 microg/kg in bottom sediments. PCB also in plants and fish. Generally, monitoring plan described.

867 Dahlgren, Robert B., and Linder, Raymond L. "Effects of Polychlorinated Biphenyls on Pheasant Reproduction, Behavior, and Survival." JOURNAL OF WILDLIFE MANAGEMENT 35 (1971): 315-19.

Experiment with pheasant chicks from hens receiving 50 mg PCB weekly. Chicks placed on a "visual cliff" for up to five minutes either jumped to the visually deep side or made no choice in significantly larger proportion than did those from untreated hens. Effects are similar to those of dieldrin on behavior.

868 Dahlgren, Robert B., et al. "Effects of Stress on Pheasants Previously Given Polychlorinated Biphenyls." JOURNAL OF WILDLIFE MANAGEMENT 36 (1972): 974-78.

Birds receiving 50 mg PCB twice a week and then alternatively starved and fed for two- to three-day intervals didn't live as long as controls, lost weight, and accumulated brain tissue residues as high as 240 ppm. Conclusion: stress causing loss of weight in birds with PCB body burden can be dangerous.

869 ______. "Storage and Excretion of Polychlorinated Biphenyls in the Pheasant." JOURNAL OF WILDLIFE MANAGEMENT 35 (1971): 823-28.

Average recovery of PCB in pheasant tissues (up to 200 ppm) was 95 percent. Absorption and storage rapid and high. Eggs important source of excretion of residue, but tissue retention very high nonetheless.

870 ______. "Residue Levels and Histopathology in Pheasants Given Polychlorinated Biphenyls." JOURNAL OF WILDLIFE MANAGEMENT 36 (1972): 524-33.

Birds dead from 210 mg PCB daily had brain residue levels of 2,500 ppm, and muscle levels of 140 ppm (averages). Brain levels of 300-400 ppm indicate PCB toxicosis. Birds that died from PCB had smaller hearts and very small spleens (due to lymphocyte depletion). Pheasants are more sensitive to PCB than to DDT.

871 Doguchi, M., and Fukano, S. "Residue Levels of Polychlorinated Terphenyls, Polychlorinated Biphenyls and DDT in Human Blood." BULLETIN OF ENVIRONMENTAL CONTAMINATION AND TOXICOLOGY 13 (1975): 57-63.

In Japan average levels in blood of PCTs, PCBs, and DDE were 5.0 ppb, 3.2 ppb, and 11.2 ppb respectively. Interestingly, PCTs have higher levels than PCBs which are produced industrially in much greater volume and which have received a good deal more attention as environmental pollutants. See references in this paper for more on PCTs.

872 Dustman, Eugene H., et al. "The Occurrence and Significance of Polychlorinated Biphenyls in the Environment." TRANSACTIONS OF THE THIRTY-SIXTH NORTH AMERICAN WILDLIFE AND NATURAL RESOURCES CONFERENCE, 1971, pp. 118-33.

Good general coverage and reference list with regard to fish and game species.

873 Fujiwara, Kunisato. "Environmental and Food Contamination with PCB's in Japan." SCIENCE OF THE TOTAL ENVIRONMENT 4 (1975): 219-47.

Excellent summary paper that should be a model for other industrial nations. Japan has had the only large scale PCB poisoning in man (1968) although at first it was called food poisoning. In 1971 Japan abruptly stopped the production, use, and importation of PCBs. No other nation has.

874 Greichus, Y.A., et al. "Effects of Polychlorinated Biphenyls in Bird Tissues and Arochlor Standards with Gas Chromatography and Mass Spectrometry." BULLETIN OF ENVIRONMENTAL CONTAMINATION AND TOXICOLOGY 11 (1974): 113-20.

All major PCBs occur in cormorant tissues and eggs; Arochlor 1254 and 1260 (commercial mixtures) also occurred in pelican fat.

875 ______. "Physiological Effects of Polychlorinated Biphenyls or a Combination of DDT, DDD, and DDE in Penned White Pelicans." ARCHIVES OF ENVIRONMENTAL CONTAMINATION AND TOXICOLOGY 3 (1975): 330-43.

Detailed accounting of various chemical effects on organs or blood chemistry. There seem to be no particular additive effects of the several types of chlorinated hydrocarbons.

876 Hansen, D.J., et al. "Arochlor 1016: Toxicity to and Uptake by Estuarine Animals." ENVIRONMENTAL RESEARCH 7 (1974): 363-73.

This PCB was tested with oysters (Crassostrea), brown shrimp (Penaeus), grass shrimp (Palaemonetas), and pinfish (Lagodon). It quickly became quite toxic to invertebrates, but prolonged exposure was required in pinfish. Tissue effects were exhibited in pinfish.

877 Haque, Rizwanul, et al. "Aqueous Solubility, Adsorption and Vapor Behavior of Polychlorinated Biphenyl Arochlor 1254." ENVIRONMENTAL SCIENCE AND TECHNOLOGY 8 (1974): 139-42.

Aqueous solubility about 56 ppm. Adsorption depends on substrate; sand adsorbs little, loams much more. Vapor loss is significantly higher from sand than from soil.

878 Harvey, G[eorge]. R., et al. "Polychlorobiphenyls in North Atlantic Ocean Water." SCIENCE 180 (1973): 643-44.

Average concentration of PCBs in North Atlantic is 20 ppt; estimated 2×10^4 metric tons of PCBs in upper 200 meters of water. Less in Sargasso Sea. Environmental half-life of PCBs much longer than DDT metabolites.

879 Hesselberg, Robert J., and Scherr, David D. "PCB's and p,p'DDE in the Blood of Cachetic Patients." BULLETIN OF ENVIRONMENTAL CONTAMINATION AND TOXICOLOGY 11 (1974): 202-5.

Patients had severe stresses, hence high metabolic fat withdrawal, as do wild species under stress. Average DDE (21 ppb) in patients' blood was higher than in normal blood. Still, no clear evidence of nervous system effects.

880 Hill, Elwood F., et al. "Polychlorinated Biphenyl Toxicity to Japanese Quail as Related to Chlorination." POULTRY SCIENCE 53 (1974): 597-604.

Chlorine percentage (in the different PCBs) correlates positively with avian toxicity, but apparently PCB toxicity is not simply a function of chlorination. Positioning of chlorine atoms on benzene rings may affect toxicity.

881 Holmes, D.C., et al. "Chlorinated Hydrocarbons in British Wildlife." NATURE 216 (1967): 227-29.

PCBs detected in wildlife. In bird livers and eggs there is often more PCB than organochlorine pesticide residues. PCBs toxic but adverse effects not known. Highest levels in terrestrial predators and in marine fish-eating birds. This paper was an early mention of PCB residue contamination, following the Swedish announcement in prior year.

882 Hutzinger, Otto, et al. THE CHEMISTRY OF PCB'S. Cleveland, Ohio: CRC Press, 1974. 269 p.

Comprehensive summary of the chemistry of the chlorobiphenyls, but does not include much on toxicology and other biological aspects. Does have section on occurrence and accumulation in the environment.

883 Hutzinger, O[tto]., et al. "Polychlorinated Biphenyls: Metabolic Behavior of Pure Isomers in Pigeons, Rats and Brook Trout." SCIENCE 178 (1972): 312-13.

Excreta examined showed alterations of some isomers in rats and pigeons; but no hydroxymetabolites were detected in excreta of brook trout.

884 Jensen, S. "Report of a New Chemical Hazard." NEW SCIENTIST (England) 36 (1966): 612.

First report on PCB residues in wild species in Sweden.

885 Kaiser, Klaus L.E., and Wong, Paul T.S. "Bacterial Degradation of Polychlorinated Biphenyls. I. Biodegradation of Some Metabolic Products from Arochlor 1242." BULLETIN OF ENVIRONMENTAL CONTAMINATION AND TOXICOLOGY 11 (1974): 291-96.

A bacterial culture, isolated from lake water, degrades fairly high concentration of commercial PCB. Some metabolites identified.

886 Koeman, J.H., et al. "Chlorinated Biphenyls in Fish, Mussels and Birds from the River Rhine and the Netherlands Coastal Area." NATURE 221 (1969): 1126-27.

PCBs are widely occurring in these regions. PCBs with fewest chlorine atoms were found in fishes but not in birds. This implies that lesser chlorinated PCBs persist less and perhaps have lesser toxicity. PCBs make detection of other chlorinated hydrocarbons difficult. DDT, DDE, and telodrin were also found.

887 Lincer, Jeffrey L., and Peakall, David B. "Metabolic Effects of Polychlorinated Biphenyls in the American Kestrel." NATURE 228 (1970): 783-84.

Estradiol breaks down to a more polar metabolite in livers of Arochlor-fed kestrels. The rate of breakdown is dose-dependent. PCBs induce liver enzymes. Action is similar to DDT and metabolites.

888 ______. "PCB Pharmacodynamics in the Ring Dove and Early Gas Chromatographic Peak Diminution." ENVIRONMENTAL POLLUTION 4 (1973): 59-68.

Ring doves were given daily dose of 10 ppm Arochlor 1254. There was linear increase of PCB in tissue until 105 days with asymptote at 80 ppm. Total analysis: in fat 797 ppm; in feces 2.27 ppm. Authors suggest that comparable levels found in starvation-stressed birds in Irish Sea in 1969 indicate that PCB was probably not the cause of death in these birds.

889 Linko, R.R., et al. "PCB Residues in Plankton and Sediment in the Southwestern Coast of Finland." BULLETIN OF ENVIRONMENTAL CONTAMINATION AND TOXICOLOGY 12 (1974): 733-38.

Plankton has relatively high amounts--0.04 to 0.75 ppm wet weight, and 4 to 77 ppm lipid weight. About the same as in fish. Amounts in sediments very low compared to plankton. PCBs do not sink to sea bottom. The source of animal contamination is the immediate surface of the sea.

890 Litterst, Charles L., and van Loon, Edward J. "Time-Course of Induction of Microsomal Enzymes Following Treatment with Polychlorinated Biphenyl." BULLETIN OF ENVIRONMENTAL CONTAMINATION AND TOXICOLOGY 11 (1974): 206-12.

Significant levels of enzyme induction within seven days. Occurs even with single oral doses. Discontinuation of dosage slowly reduces induction to normal in about ten days.

891 Masuda, Yoshita, et al. "Comparisons of Polychlorinated Biphenyls in Yusho Patients and Ordinary Persons." BULLETIN OF ENVIRONMENTAL CONTAMINATION AND TOXICOLOGY 11 (1974): 213-16.

PCB hazard known since 1966 (Sweden). There was a mass outbreak of human poisoning (called "Yusho") by PCB-contaminated rice oil in Japan in 1968. Average value of PCBs slightly higher in Yusho patients. Milk in one Yusho patient contained 0.06 ppm.

892 Maugh, Thomas H. II. "DDT: An Unrecognized Source of PCB." SCIENCE 180 (1973): 578-79.

Reported that DDT vapor can be converted to PCBs by irradiation with ultraviolet light of the same wavelengths present in sunlight in the lower atmosphere. DDT and DDE are generally inert in sunlight, but not true of their vapors. Effect is substitution of one environmental contaminant with another.

893 Metcalf, Robert L., et al. "Laboratory Model Ecosystem Studies of the Degradation and Fate of Radiolabelled Tri-, Tetra-, Pentachlorobiphenyl Compared with DDE." ARCHIVES OF ENVIRONMENTAL CONTAMINATION AND TOXICOLOGY 3 (1975): 151-65.

"Tri-" degraded much more rapidly than others. "Penta-" about the same as DDE. Linear relationship between lipid/water partition water solubility, and ecological magnification. No evidence of conversion of DDE to PCB. Good reference list.

894 Morgan, Janet R. "Effects of Arochlor 1242 (a Polychlorinated Biphenyl) and DDT on Cultures of an Alga, Protozoan, Daphnid, Ostracod, and Guppy." BULLETIN OF ENVIRONMENTAL CONTAMINATION AND TOXICOLOGY 8 (1972): 129-37.

Arochlor (but not DDT) inhibited growth of *Chlamydomonas* (alga) and concentrated over medium in the alga. *Daphnia* and the Cladoceran (*Cypridopsis*) highly susceptible to both Arochlor and DDT. Young guppies died in medium of 2 ppm Arochlor, a lower toxicity than p,p'-DDT. Clearly, Arochlor can concentrate in basal members of food chain.

895 Mosser, Jerry L., et al. "Interactions of PCBs, DDT and DDE in a Marine Diatom." BULLETIN OF ENVIRONMENTAL CONTAMINATION AND TOXICOLOGY 12 (1974): 665-68.

Authors suggest that organochlorines and PCBs may interact for greater toxicity than shown singly.

896 ______. "Polychlorinated Biphenyls and DDT Alter Species Composition in Mixed Cultures of Algae." SCIENCE 176 (1972): 533-35.

Each chemical was added to mixed cultures of a marine diatom and a marine green alga and produced similar effects. Diatoms grew faster in controls; dominance diminished in treated cultures, even at concentrations of chlorinated hydrocarbons that had no effect in pure cultures. Stable pollutants could disrupt species composition of phytoplankton communities, thereby affecting whole ecosystems. (There is no field evidence of this, however.)

897 Nimmo, D.R., et al. "Accumulation of Arochlor ®1254 in Grass Shrimp (*Palaemonetes pugio*) in Laboratory and Field Exposures." BULLETIN OF ENVIRONMENTAL CONTAMINATION AND TOXICOLOGY 11 (1974): 303-8.

Study of a shrimp that is estuarine throughout its life cycle; shown to be bioaccumulator of residues. But lesser exposure than species with pelagic phase in life cycle. Shrimp exposed to PCB-contaminated sediments accumulated 0.42 mg/kg after three months without significant mortality. Higher accumulation and mortality in laboratory tests.

898 _____. "Toxicity of Arochlor 1254 and Its Physiological Activity in Several Estuarine Organisms." ARCHIVES OF ENVIRONMENTAL CONTAMINATION AND TOXICOLOGY 3 (1975): 22-39.

High concentrations in Florida estuary; present in all components. Toxicity tests indicate organisms sensitive at ppb levels; a ciliate protozoan, two genera of shrimp, and killifish affected at or near 1.0 ppb. Shrimp concentrated PCB in tissue from as low as 0.04 ppb in water. Pathological changes in oysters, in spot (fish), and in shrimp. Paper combines field and laboratory data well.

899 Nisbet, Ian C.T., and Sarofim, Adel F. "Rates and Routes of Transport of PCB's in the Environment." ENVIRONMENTAL HEALTH PERSPECTIVES 1 (1972): 21-38.

Identifies sources, amounts, and problems connected with PCBs in all parts of the environment. Note particularly bioconcentration model and DDT/PCB residue ratios.

900 Panel on Hazardous Trace Substances. "PCBs-- Environmental Impact." ENVIRONMENTAL RESEARCH 5 (1972): 249-362.

Excellent detailed review of PCBs. Much original data. Tables in text and extensive bibliography.

901 Peakall, David B. "PCB's and Their Environmental Effects." CRITICAL REVIEWS IN ENVIRONMENTAL CONTROLS 6 (1975): 469-508.

The most important review of the subject. Two hundred fifty-one references. All aspects of PCBs discussed.

902 _____. "Polychlorinated Biphenyls: Occurrence and Biological Effects." RESIDUE REVIEWS 44 (1972): 1-21.

PCBs now widespread, paralleling DDT in concentration. Very stable; will persist for many years. Chief manufacturer has restricted production. PCBs embryotoxic in 10 ppm range.

903 Risebrough, Robert W., et al. "PCB Residues in Atlantic Zooplankton." BULLETIN OF ENVIRONMENTAL CONTAMINATION AND TOXICOLOGY 8 (1972): 345-55.

In many areas of the Atlantic Ocea PCB residues exceed those of DDT. PCB in lipids of zooplankters from the continental shelf ranged from 2.4 to 260 ppm. From shelf and slope (wet weight) levels were about 1.5 ppm and 0.15 ppm respectively.

904 ______. "Polychlorinated Biphenyls in the Global Ecosystem." NATURE 220 (1968): 1098-1102.

This was the first announcement that these persistent waste products are distributed worldwide, survive in environment and tissues for long periods, and accumulate in food chains.

905 Rote, James W., and Morris, William J. "Use of Isotopic Abundance Ratios in Identification of Polychlorinated Biphenyls by Mass Spectrometry." JOURNAL OF THE ASSOCIATION OF OFFICIAL ANALYTICAL CHEMISTS 56 (1973): 188-99.

This method provides unambiguous identification of PCBs in the presence of other substances.

906 Rote, J[ames]. W., and Murphy, Philip G. "A Method for the Quantitation of Polychlorinated Biphenyl (PCB) Isomers." BULLETIN OF ENVIRONMENTAL CONTAMINATION AND TOXICOLOGY 6 (1971): 377-84.

PCBs are complex mixtures of isomers having different degrees of chlorination. Peaks on chromatograms depend on chlorination in varieties of Arochlors.

907 Stickel, Lucille F. "Biological Data on PCBs in Animals Other Than Man." In POLYCHLORINATED BIPHENYLS AND THE ENVIRONMENT, pp. 158-72. Interdepartmental Task Force on PCBs. Springfield, Va.: National Technical Information Service, for U.S. Department of Commerce, 1972.

A summary review of PCB effects on wild species of animals. Comparisons made with chlorinated hydrocarbon pesticides. Author is one of several representatives of federal agencies contributing to this publication.

908 Sugiura, K., et al. "Accumulation and Excretion of PCBs in the Mouse." CHEMOSPHERE 3 (1975): 181-87.

Detailed chemistry indicating that residues of PCB depend on an adsorption-diffusion mechanism in adipose tissue, as do other chlorinated hydrocarbons. Seemingly, PCBs with fewer chlorine atoms are more rapidly metabolized and excreted.

909 Sundstrom, Gonar. "Polychlorinated Biphenyls. V. $^{+}$Synthesis of ^{14}C-Labelled 2,2',4,4',5,5'--Hexachlorobiphenyl, a Major Component in PCB-Mixtures." BULLETIN OF ENVIRONMENTAL CONTAMINATION

AND TOXICOLOGY 11 (1974): 39-44.

Report of synthesis of PCB component material.

910 Thompson, Neal P., et al. "Polychlorinated Biphenyls and p,p'-DDE in Green Turtle Eggs from Ascension Island, South Atlantic Ocean." BULLETIN OF ENVIRONMENTAL CONTAMINATION AND TOXICOLOGY 11 (1974): 399-406.

Multiple PCBs found. All residues low. Island remote. No effects shown including in eggs brought to Florida for hatching.

911 Tucker, E.S., et al. "Migration of Polychlorinated Biphenyls in Soil Induced by Percolating Water." BULLETIN OF ENVIRONMENTAL CONTAMINATION AND TOXICOLOGY 13 (1975): 86-93.

PCBs are not readily leached from soil by percolating water, and only the less chlorinated, more degradable, homologs are leached from soil. Problem is disposal of Arochlor-impregnated capacitors. Authors conclude that landfills are only a minor source of PCB environmental contamination.

912 Tumasonis, Casimir Franco, et al. "PCB Levels in Egg Yolks Associated with Embryonic Mortality and Deformity of Hatched Chicks." BULLETIN OF ENVIRONMENTAL CONTAMINATION AND TOXICOLOGY 1 (1973): 312-24.

Hens were exposed for six weeks to Arochlor 1254. Decreased egg production but no effects on egg weights or fertility. Leg, toe, and neck deformities occurred in chicks hatched from eggs in which PCB levels were 10-15 ppm or more. Liver is a target for PCB action. Good bibliography.

913 U.S. Department of Health, Education, and Welfare. "Perspective on PCBs." ENVIRONMENTAL HEALTH PERSPECTIVES, Experimental Issue no. 1 (1972): entire issue (185 p.).

Thirty contributions on a wide range of subjects. Published results of a conference held in 1971. Topics include toxicity to vertebrates, transport means in the environment, and effects in aquatic systems.

914 Yoshida, T., et al. "Distribution of [^{14}C] PCB's in Carp." AMBIO 2 (1973): 111-13.

Study relative to known bioconcentration in fishes and fish-eating birds. Which organs in fish concentrate? Autoradiography shows localization mainly in gall bladder, fat, and hepatopancreas. Highest in gall bladder and in fatty tissue of skulls.

915 Zepp, R.L., Jr., et al. "Reduction of Pentobarbital-induced Sleeping Times in PCB-treated Cottontail Rabbits." BULLETIN OF ENVIRONMENTAL CONTAMINATION AND TOXICOLOGY 12 (1974): 518-21.

Ten ppm Arochlor 1254 significantly reduced sleeping times of both sexes. Relates to hepatic microsomal enzyme activity.

916 Zinck, M.E., and Addison, R.F. "The Fate of 2-, 3-, and 4-, Chlorobiphenyl Following Intravenous Administration to the Thorny Skate (Raja radiata) and the Winter Skate (Raja ocellata)." ARCHIVES OF ENVIRONMENTAL CONTAMINATION AND TOXICOLOGY 2 (1974): 52-61.

Isomers of PCB with low chlorine content. All accumulated in tissues. The 3- isomer cleared most rapidly in plasma and from tissue. Considerable variance in residue levels from fish to fish.

917 Zitko, V., et al. "Distribution of PCB and p,p'-DDE Residues in Atlantic Herring (Clupea harengus harengus) and Yellow Perch (Perca flavescens) in Eastern Canada--1972." PESTICIDES MONITORING JOURNAL 8 (1974): 105-9.

Detailed study intended to determine temporal trends in contaminants. Concluded impossible on year-to-year basis because of extreme variability.

Section 12

PLANT AND ANIMAL TOXINS AND VENOMS

Toxic materials do exist commonly as natural products; plant and animal toxins sometimes can be used to man's advantage in pest or disease control. Occasionally the determination of the cause of a toxic effect is confused by naturally occurring compounds, whether animal, vegetable, or mineral. Beyond the clear interests of the environmentalist and toxicologist, naturally occurring toxicants in living organisms hold a general fascination. From our standpoint they can be viewed as comparable to the pheromones--distinctive chemical products manufactured by living organisms and serving very specific biological ends.

918 Bridges, W[illiam]. R., and Cope, Oliver B. "The Relative Toxicities of Similar Formulations of Pyrethrum and Rotenone to Fish and Immature Stoneflies." PYRETHRUM POST 8 (1965): 3-5.

Rotenone formulation is 1.8 to 3.2 times more toxic to fish than pyrethrum, but pyrethrum is 141 to 290 times more toxic to stoneflies.

919 Brooks, Ivan C., and Price, Robert W. "Studies on the Chronic Toxicity of Pro-Noxfish, a Proprietary Synergized Rotenone Fish Toxicant." TOXICOLOGY AND APPLIED PHARMACOLOGY 3 (1961): 49-56.

Neither the rotenone nor sulfoxide ingredients seemed to affect rats when fed in diet. Mammalian safety implied.

920 Burdick, George E., et al. "Toxicity of Emulsifiable Rotenone to Various Species of Fish." NEW YORK FISH AND GAME JOURNAL 2 (1955): 36-67.

Detailed reference appearing at time when rotenone was beginning to be widely used as a toxin in fisheries management.

921 Casida, John E., et al., eds. PYRETHRUM: THE NATURAL INSECTICIDE. New York: Academic Press, 1973. xvii, 329 p.

Twenty-three contributors on many aspects--history, production, chemistry, testing methods, toxicity, mutagenic effects, mode of action, and range of use.

922 Culvenor, C.C.J. "Toxic Plants--A Reevaluation." SEARCH (Australia) 1 (1970): 103-10.

Excellent popular review of the taxonomy and chemical character of toxic plants. Summary tables.

923 Daly, John W., and Myers, Charles W. "Toxicity of Panamanian Poison Frogs (Dendrobates): Some Biological and Chemical Aspects." SCIENCE 156 (1967): 970-73.

Some amphibians are extremely toxic, and show many biological adaptations to the presence of toxins. This paper extremely useful; particularly useful is a table illustrating comparative toxicity of chemicals found in amphibians. For example, lethal dose to mice in many such poisons lower than in strychnine and sodium cyanide.

924 Habermehl, Gerhard G. "Venoms of Amphibia." In CHEMICAL ZOOLOGY, edited by Marcel Florkin and Bradley T. Scheer, vol. 9, pp. 161-83. New York: Academic Press, 1974.

Summary of all aspects of venom production in frogs, toads, and salamanders. Naturally produced venoms differ in the major groups and are generally protective secretions.

925 Halstead, Bruce W., ed. POISONOUS AND VENOMOUS MARINE ANIMALS OF THE WORLD. 3 vols. Sponsored by U.S. Department of Defense. Washington, D.C.: Government Printing Office, 1964-70. xxxv, 994 p.; xxvi, 1,070 p.; xxxv, 1,006 p.

Monumental reference. Volume 1 is PROTOZOANS THROUGH FLATWORMS: volumes 2 and 3 are VERTEBRATES. Abundant illustration, tables, referencing, and exposition.

926 Humm, Harold J., and Lane Charles E., eds. BIOACTIVE COMPOUNDS FROM THE SEA. New York: Marcel Dekker, 1974. xiii, 251 p.

Thirteen chapters by different authors describing a wide variety of chemicals including toxins from echinoderms, dinoflagellates, fishes, and sea snakes. Descriptions are technical and include toxicology. Extensive referencing given in each chapter.

927 Keegan, Hugh L., and MacFarlane, W.V. VENOMOUS AND POISONOUS ANIMALS AND NOXIOUS PLANTS OF THE PACIFIC REGION. New York: Macmillan Co., 1963. xi, 456 p.

Remarkably varied coverage by individual contributors. Authoritative reference.

928 Kingsbury, John M. DEADLY HARVEST: A GUIDE TO COMMON POISONOUS PLANTS. New York: Holt, Rinehart and Winston, 1965. 128 p.

Excellent brief survey with drawings of common poisonous plants. Commonsensical and interesting treatment in text.

929 ______. POISONOUS PLANTS OF THE UNITED STATES AND CANADA. Englewood Cliffs, N.J.: Prentice-Hall, 1964. xiii, 626 p.

Compendium; taxonomically organized. Eighty-one-page bibliography.

930 Kiser, R.W., et al. "The Effect of Rotenone on Zooplankton Population in Freshwater Lakes." TRANSACTIONS OF THE AMERICAN FISHERIES SOCIETY 92 (1963): 17-24.

Cladocerans and copepods are severely affected by rotenone normally used in fish management. Effects are most disturbing; several months required for population recovery.

931 Manske, R.H.F., ed. THE ALKALOIDS. Vol. 14. New York: Academic Press, 1973. xvii, 610 p.

Latest in series of volumes (vol. 1, 1950), the first five volumes of which were edited by R.H.F. Manske. Format and title the same for all volumes. Volume 14 contains twelve chapters by individual contributors. Lists authors and titles in prior volumes in the series.

932 Pelletier, S.W., ed. CHEMISTRY OF THE ALKALOIDS. New York: Van Nostrand Reinhold Co., 1970. xx, 795 p.

There are many thousands of alkaloid compounds. All are produced by plants. Twenty-two chapters in this book by different authors cover topics ranging from general information to use in plant systematics.

933 Russell, Findlay E. MARINE TOXINS AND VENOMOUS AND POISONOUS MARINE ANIMALS. London: Academic Press, 1965. Distributed by T.F.H. Publications, Neptune City, N.J. 176 p.

Eight chapters describe the some 1,000 species of toxic marine organisms. Range is from one-celled organisms to fishes. Comprehensive; long reference list. Chemistry and toxicology included.

934 Russell, Findlay E., and Brodie, Arnold F. "Venoms of Reptiles." In CHEMICAL ZOOLOGY, edited by Marcel Florkin and Bradley T. Scheer, vol. 9, pp. 449-78. New York: Academic Press, 1974.

Summary of venom characteristics in reptiles: of 3,500 species of snakes, less than 400 are hazardous to man. Chemical compositions of these are tabulated. Many references.

935 TOXICON. Oxford, Engl.: Pergamon Press, 1963--. Bimonthly. Approximately 540 pages per year.

International journal devoted to exchange of knowledge on poisons derived from animals, plants, and microorganisms.

Part IV

SPECIAL ASPECTS OF ENVIRONMENTAL TOXICOLOGY

The need to control chemical usage for the best human interests in production of foods and fibers and in disease prevention has long been recognized. But one cannot know the effects or occurrence of chemical pollutants without checking for them. Hence agency programs have been established to monitor residual amounts and to check indicator species as warning signals. These warning systems are for predictive human protection.

Animals, too, have adopted protective systems. But they do so in a rapid evolutionary manner--by selecting out those individuals in a population most resistant to the presence of toxic chemicals and breeding genetically resistant offspring. Arthropods, fishes, and mammals have evolved such populations.

Given the abundant problems created by chemicals in the environment, we must continuously seek methods of alleviating them. Currently, we do so as well as possible in several ways. The better known are biological and integrated pest control; these and other methods are the subject of section 16.

Section 13

LEGISLATION AND REGULATION

This sourcebook does not emphasize legislative and regulatory aspects of toxic pollutants. ENVIRONMENTAL LAW, another book in the Man and the Environment series, describes such matters in detail. For convenience, however, I have listed and described some laws, agencies, and literature sources that relate directly to environmental toxicology. This list is only a sampling of a very large body of laws, regulations, and codes. I have deliberately included a few sources from countries other than the United States to show how regulatory practices differ with cultural and ecological conditions. In general, state, national, and international standardization of regulatory mechanisms has not been achieved. The goal is salutary, however, and progress continues to be made toward larger comprehension and its necessary codification.

936 FEDERAL ACTS RELATING TO POLLUTION

1. Rivers and Harbors Act of 1899
2. Federal Insecticide Act of 1910
3. Federal Food, Drug, and Cosmetic Act of 1938
4. Federal Insecticide, Fungicide, and Rodenticide Act of 1947
5. Water Pollution Control Act of 1948
6. Saline Water Conservation Act of 1952, amended 1961
7. Multiple Use-Sustained Yield Act of 1960
8. Oil Pollution Act of 1961
9. Water Resources Research Act of 1964
10. Wilderness Act of 1964
11. Solid Waste Disposal Act of 1965
12. Water Resources Planning Act of 1965
13. Clean Water Restoration Act of 1966
14. Marine Resources and Engineering Development Act of 1966
15. Clean Air Act of 1967
16. National Environmental Policy Act of 1969
17. Environmental Education Act of 1970
18. Creation of Environmental Protection Agency, 1970
19. Water Quality Improvement Act of 1970
20. Endangered Species Act of 1971

RESPONSIBILITIES OF CERTAIN ACTS AND AGENCIES

937 Federal Insecticide Act of 1910

Basically intended to protect the farmer from adulterated or misbranded products. Before that year, there was no buyer protection of any sort.

938 Federal Insecticide, Fungicide, and Rodenticide Act of 1947 (FIFRA)

Added a new concept, that of placing the burden of proof of efficacy and safety on the manufacturer. Proof had to be furnished and labels written accurately. Amended by the Federal Environmental Pesticide Control Act of 1972 (PL 92-516), known as FEPCA. Includes a four-year "adaptive" period in which all states are expected to modify to minimal federal standards. States may, however, adopt more restrictive standards. This act, for the first time, recognized the environment as a quality to be protected.

939 Federal Food, Drug, and Cosmetic Act of 1938

Administered by the Food and Drug Administration; concern originally with immediate toxicity. FDA in 1948 began to establish safe tolerance levels of residues in foods of human beings and domestic animals. Levels were not established higher than those registered as approved by U.S. Department of Agriculture.

The so-called Miller Amendment (July 22, 1954; Public Law 518; 83d Cong.) formalized arrangements whereby USDA registered only pesticide uses resulting in no residues on agricultural products or residues declared safe by the FDA. Registrants had to submit data to both USDA and FDA, and tolerance levels were set 100 times over known, measurable effects on warm-blooded animals.

The related Delaney Amendment (1958) has the special purpose of dictating that no chemical known to be carcinogenic may be added in foodstuffs, cosmetics, or drugs. Unfortunately, it dealt only with intentional additives.

940 Environmental Protection Agency (EPA)

Created December 2, 1970, under a presidential reorganizational plan that brought major environmental control programs under one agency. Included were pesticide regulatory programs, water and air quality standards, nuclear radiation controls, tolerance setting, and monitoring programs. Occupational health considerations were left under jurisdiction of the Department of Health, Education, and Welfare. For some years a more encompassing act--the Toxic Substances Control Act--has been proposed to Congress. The bill is currently (1976) being more strongly argued than previously. Its nature is simple: chemicals have many predictable features, and their biological and human consequences should be central considerations in environmental control, not their practical use or origins.

EPA would administer the act and also administers NEPA--the National Environmental Policy Act of 1969. This act controls federal agency activities and was the first law requiring environmental impact statements.

Other Federal Agencies

941 The Department of Transportation has regulations, more stringently codified in 1974, relating to interstate transport of hazardous materials.

942 The Federal Aviation Administration (FAA) regulates the dispensing of pesticides by aircraft. Authority for this purpose began in 1966 but has been since amended more specifically. Any aerial applicator must use methods that are not inconsistent with directions for use on the federally registered label.

943 The Department of Agriculture now controls residues largely through its Consumer and Marketing Services. Sources of contamination are traced, and follow-up inspections ensure satisfactory future marketable products.

LITERATURE SOURCES: GENERAL REFERENCES

944 Barros, James, and Johnston, Douglas M. THE INTERNATIONAL LAW OF POLLUTION. New York: The Free Press, 1974. 476 p.

Chiefly contains proposed laws and interpretations. Particularly useful are parts on definitions, pollutant identification, and criterion of pollution assessment and control.

945 Brecher, Joseph J., and Nestle, Manuel E. ENVIRONMENTAL LAW HANDBOOK. Berkeley: California Continuing Education at the Bar, n.d. 343 p. Paper.

Excellent compendium and reference. Many ecological and toxicological references. Long reference list on legal precedents.

946 ECOLOGY LAW QUARTERLY. Berkeley: School of Law, University of California, Berkeley, 1970--. Quarterly. Approx. 800 pages per year.

Reviews of laws and governmental regulations that concern ecological matters. The scope is national; most papers are prepared by advanced students in Boalt School of Law.

947 Epstein, Samuel S., and Grundy, R.D., eds. COMESTICS AND DRUGS, PESTICIDES AND FOOD ADDITIVES. The Legislation of Product Safety: Consumer Health and Product Hazards, vol. 2. Cambridge: Massachusetts Institute of Technology Press, 1974. 342 p.

Part of larger series but only this volume is relevant. See particularly "Toxic Substances," pages 101-69.

948 Fine, S.D. FINAL ENVIRONMENTAL IMPACT STATEMENT: RULE MAKING ON POLYCHLORINATED BIPHENYLS. Washington, D.C.: Food and Drug Administration, 1972. 29 p.

Registration limitations and procedures for the PCBs.

949 Hazeltine, William E. THE LEGISLATIVE HISTORY AND MEANING OF THE FEDERAL INSECTICIDE, FUNGICIDE AND RODENTICIDE ACT AS AMENDED, 1972. Private printing, 1975. Available from William Hazeltine, 26 Rosita Way, Oroville, Calif. 95656.

Contains twenty-nine chapters, separately paginated. Laws, congressional hearings, and commentary by author. Detailed. Author maintains that EPA does not improve accuracy and speed of pesticide regulation, but actually slows regulation.

950 Headley, J[oseph]. C[harles]., and Lewis, J.N. THE PESTICIDE PROBLEM: AN ECONOMIC APPROACH TO PUBLIC POLICY. Washington, D.C.: Resources for the Future, 1967. xvii, 141 p. Paper.

Basically, an attempt to put competition between man and other organisms--and the use of pesticides to diminish competition--in an understandable economic context. The sponsoring foundation, Resources for the Future, is concerned with all aspects of public policy regarding the use of natural resources. This study examines the pesticide problem on an economic basis.

951 Henkin, Harmon, et al. THE ENVIRONMENT, THE ESTABLISHMENT, AND THE LAW. Boston: Houghton Mifflin Co., 1971. vii, 223 p. Paper.

Testimony and commentary derived from public hearings in Madison, Wisconsin, in 1968-69 on whether DDT was to be ruled an environmental pollutant according to statutes of the state of Wisconsin. Provocative, sometimes heated testimony, but still a landmark event and worth referring to by those unfamiliar with legal hearing procedures on scientific matters.

952 Krier, J.G. "Environmental Law and Its Administration." In ENVIRONMENT--RESOURCES, POLLUTION AND SOCIETY, edited by William W. Murdoch, pp. 413-36. 2d ed. Sunderland, Mass.: Sinauer Associates, 1975.

Very useful general treatment of legal aspects of environmental problems by a lawyer. Particularly good for environmentalists with no knowledge of regulations, injunctions, court procedures, and other legal matters.

953 Mylroie, Gerald R., and Towne, Karen. CALIFORNIA ENVIRONMENTAL LAW: A GUIDE. 4th ed. Claremont: Center for California Public Affairs, 1974. 188 p. Paper. Available from publishers, 226 West Foothill Boulevard, Claremont, Calif. 91711.

Includes legislative summations affecting environmental quality, pesticides, water quality, pollution, and wildlife.

954 Organization for Economic Cooperation and Development (OECD). POLYCHLORINATED BIPHENYLS--THEIR USE AND CONTROL. Council decision on protection of the environment by control of polychlorinated biphenyls. Paris: 1973. 44 p.

Suggests measures to reduce uncontrolled releases of PCBs into the environment. Specific industrial uses are acceptable. (Production and import of PCBs was stopped in Japan in 1972, but in no other country.)

955 Udell, Gilman G., comp. LAWS RELATING TO AGRICULTURE. Washington, D.C.: Government Printing Office, 1971. xv, 840 p.

Texts of all laws affecting the U.S. Department of Agriculture, from the original 1884 bill to the Food Stamps program.

956 U.S. Congress, House. Committee on Public Works. LAWS OF THE UNITED STATES RELATING TO WATER POLLUTION CONTROL AND ENVIRONMENTAL QUALITY. Washington, D.C.: Government Printing Office, 1970. 265 p.

Complete federal legislation to 1970.

957 U.S. Congress, House. Joint Committee: House Committee on Interstate and Foreign Commerce; Senate Committee on Labor and Public Welfare. COMPILATION OF SELECTED PUBLIC HEALTH LAWS. INCLUDING PARTICULARLY THE PUBLIC HEALTH SERVICE ACT, THE CLEAN AIR ACT, THE SOLID WASTE DISPOSAL ACT, THE MENTAL RETARDATION FACILITIES AND COMMUNITY MENTAL HEALTH CENTERS CONSTRUCTION ACT, THE NARCOTIC ADDICT REHABILITATION ACT, THE FEDERAL FOOD, DRUG AND COSMETIC ACT, THE FEDERAL HAZARDOUS SUBSTANCES ACT, AND THE FAIR PACKAGING AND LABELING ACT. Vol. 2. 93d Cong., 1st sess., 1973.

This volume includes texts of thirteen food, drug, and related laws beginning with the Federal Food, Drug, and Cosmetic Act (1938). Not all relate directly to environmental problems.

958 U.S. Congress, House. Subcommittee of Committee on Government Operations. MERCURY POLLUTION AND ENFORCEMENT OF THE REFUSE ACT OF 1899 (PART 3): HEARINGS. 92d Cong., 2d sess., 21 December 1972, pp. 1271-482.

Testimony and affidavits relating to significance and legislative control of mercury pollution.

959 U.S. Congress, Senate. Environmental Subcommittee of Committee on Commerce. TOXIC SUBSTANCES CONTROL ACT OF 1973: HEARINGS ON S. 888, 93d Cong., 1st sess., 1973.

Testimony and evidence. This bill in same form has been initiated and delayed for five years. It is highly controversial. In 1976 the bill is being strongly backed again by the Environmental Protection Agency.

960 U.S. Secretary of the Interior. THE NATIONAL ESTUARINE POLLUTION STUDY. Report of the Secretary pursuant to PL 89-753 (The Clean Water Restoration Act of 1966), 91st Cong., 2d sess. Doc. no. 91-58, 1970. ix, 633 p.

Seven parts with chapters and sections within each. Toxic materials presented in many sections. Recommendations, references, legislation. Major analysis of the Clean Waters bill and its intentions.

961 World Health Organization. CONTROL OF PESTICIDES--A SURVEY OF EXISTING LEGISLATION. Geneva: 1970. 150 p.

Covers definition of the term "pesticide"; authorities involved in legislating the use, sale, or manufacture of pesticides; systems of approval or registration of pesticides; control of manufacture and supply; labeling; occupational health; aerial application; residues in foodstuffs; and other aspects. Specific legislation of eleven countries, including the United States, is discussed.

962 Yannacone, Victor J., Jr., et al. ENVIRONMENTAL RIGHTS AND REMEDIES. 2 vols. Rochester, N.Y.: Lawyers Cooperative Publishing Co.; San Francisco: Bancroft, Whitney and Co., 1972. xi, 655 p; vi, 704 p.

Unusually lucid legal reference dealing with environmental problems including intruded chemicals.

963 Younger, Evelle J. "Primer on Environmental Law in California." 3d ed. Compiled for general use as source material by the Attorney General's Environmental Unit. Sacramento: State of California, 1974. 56 p. Mimeo.

Laws and interpretations regarding pollution, wildlife, and endangered species.

LITERATURE SOURCES: JOURNAL ARTICLES

964 Benes, V., and Cerna, V. "Tolerances of Pesticide Residues in Czechoslovakia." RESIDUE REVIEWS 33 (1970): 75-85.

Residue tolerance thresholds were established in 1965. Emphasis on occupational and consumer hazards. Compromise with needs of agricultural pest control needs. Several field illustrations.

965 Bevenue, Arthur, and Kawano, Yoshihiko. "Pesticides, Pesticide Residues, Tolerances, and the Law (U.S.A.)." RESIDUE REVIEWS 35 (1971): 103-49.

Descriptions of operative legislation and agency formations and actions influencing pesticide regulation. The report centers on the Federal Insecticide, Fungicide, and Rodenticide Act and the Federal Food, Drug, and Cosmetic Act, and two important "incidents"--controversy regarding DDT and 2,4-D.

966 Deck, Errett. "Federal and State Pesticide Regulations and Legislation." ANNUAL REVIEW OF ENTOMOLOGY 20 (1975): 119-31.

Review of government enactments to control pesticidal use; concentrates on federal regulations. Registration, applicators, consumers, and environment are considered. The report is still controversial in parts, but in general offers an amazingly successful scheme of control. The article is "biased" in the sense that the author is a regulator and argues for success of regulatory system. His arguments are convincing, however.

967 de Pietri-Tonelli, P. "The Regulation of Pesticides in Italy." RESIDUE REVIEWS 27 (1969): 1-18.

Detailed account of Italian legislation, registration, inspection, and advisory systems. Italian regulation appears similar to U.S. system.

968 Dunning, Harrison C. "Pests, Poisons, and the Living Law: The Control of Pesticides in California's Imperial Valley." ECOLOGY LAW QUARTERLY 2 (1972): 633-93.

Cost and hazards dictate that no single method of pest control is effective; stresses integrated control. Heavily footnoted with regulatory, legal, and biological information.

969 Fukunaga, K., and Tsukano, Y. "Pesticide Regulations and Residue Problems in Japan." RESIDUE REVIEWS 26 (1969): 1-16.

Three laws govern pesticide use: (1) the Agricultural Chemicals Control Law; (2) the Poisonous and Deleterious Substance Control Law; and (3) the Food Sanitation Law. Present goals are to eliminate all highly toxic chemicals but not to decrease yields in agricultural crops.

970 Locatelli, M., and DeFalco, G. "The Regulation of Pesticides in Argentina." RESIDUE REVIEWS 44 (1972): 39-64.

Pest regulation dates officially from 1888. Single national body approves, rejects, or tests materials. Threshold limits are set and labeling is controlled. Operation of law concerns commercial sales and consumer hazards.

971 Lu, Frank C. "Toxicological Evaluation of Food Additives and Pesticide Residues and Their 'Acceptable Daily Intakes' for Man: The Role of WHO in Conjunction with FAD." RESIDUE REVIEWS 45 (1973): 81-93.

Globally, there is steady increase in intentional additives and pesticide residues. The United Nations depends on expert committees to evaluate problems. Lu recommends internationally acceptable regulations.

972 Markkula, M. "Regulation of Pesticides in Finland." RESIDUE REVIEWS 48 (1973): 117-39.

Reports extent of use and manner of regulation in Finland. Regulation is based on inclusive pesticide act of 1969. Any chemical harming man or nature can be banned. Sales permitted only with five-year license.

973 Silva Fernandes, A.M.S. "Pesticide Legislation and Residue Problems in Portugal." RESIDUE REVIEWS 35 (1971): 29-47.

Sales licensed according to four classes of chemicals, varying with hazard. Classes (1-4) are indicated on labels. Summary of residue work in country through 1969.

974 Stenersen, Jorgen. "Pesticides for Plant Protection in Norway: Legislation Use and Residues." RESIDUE REVIEWS 42 (1972): 91-102.

Reviews pesticide use legislation, and also residues in wildlife and environment. No major pollution problem anticipated.

975 Stobieck, Tadeusz. "Pesticide Regulations and Residue Problems in Poland." RESIDUE REVIEWS 33 (1970): 1-4.

Discusses Polish agriculture in relation to the needed registration and regulation of pesticides. Thus far there is no great problem with residues, because of moderate use of pesticides.

976 Sugawara, Ken. "Marine Pollution in Japan and How to Control It." CHEMOSPHERE 6 (1974): 281-86.

General statement that includes recent official actions of the Japanese government regarding oceanic pollution.

977 Van Tiel, N. "Principles and Problems of Pesticides Regulation." PESTICIDES SCIENCE 6 (1975): 189-97.

A good, general paper outlining regulatory problems more particularly as they relate to control and advice under U.N. organizations. Basic problem is how to employ more than one tolerance figure and how to "escape" national use regulations which often do not conform to suggested international standards.

978 Waitt, Arthur W. "Pesticide Legislation and Industry." PESTICIDES SCIENCE 6 (1975): 199-208.

The methods by which improved technology of pesticides might be regulated both nationally and internationally. A business point of view.

979 Wiese, I.H., and Bot, J. "Pesticide Regulation in South Africa." RESIDUE REVIEWS 35 (1971): 49-63.

South Africa has essentially compulsory registration, with a single control. Other departments have indirect control. Fairly complex set of regulations.

980 Wissa, Samir. "Pesticide Regulation in Egypt." RESIDUE REVIEWS 42 (1972): 165-72.

History of operative laws and proceedings from 1921 to 1967. Registration, marketing, packaging, and surveillance phases.

Section 14

MONITORING THE ENVIRONMENT

Only governmental agencies can support and administer the needed network of monitoring stations on a proper scale. Universities and large corporations provide some original information and cooperate with state and federal agencies. Section 13 identified operative legislation. The brunt and the task of this operating program is borne by the Environmental Protection Agency, the U.S. Public Health Service, the U.S. Fish and Wildlife Service, and state and federal consumer protection agencies under departments of agriculture.

981 Cothern, C.R. DETERMINATION OF TRACE METAL POLLUTANTS IN WATER RESOURCES AND STREAM SEDIMENTS. Washington, D.C.: Department of Interior, Office of Water Resources Research, 1974. Distributed by National Technical Information Service, U.S. Department of Commerce, Springfield, Va. 22151. 182 p.

Technical report on physical aspects of monitoring trace metals and radioisotopes.

982 U.S. Department of Health, Education, and Welfare. MAN'S HEALTH AND THE ENVIRONMENT--SOME RESEARCH NEEDS. Washington, D.C.: Government Printing Office, 1970. xix, 258 p.

Chiefly health related but useful for its range of coverage and suggestions on research, surveillance, monitoring, estimation of hazard, and so on.

983 Dieter, Michael P. "Further Studies on the Use of Enzyme Profiles to Monitor Residue Accumulation in Wildlife: Plasma Enzymes in Starlings Fed Graded Concentrations of Morsodren, DDE, Arochlor 1254, and Malathion." ARCHIVES OF ENVIRONMENTAL CONTAMINATION AND TOXICOLOGY 3 (1975): 142-50.

Enzymatic parameters easily identify actions of chemical classes. Different residue levels in tissues also described.

984 Environmental Protection Agency. PROCEEDINGS OF SEMINAR ON METHODOLOGY FOR MONITORING THE MARINE ENVIRONMENT. Washington, D.C.: Office of Research and Development, EPA, 1974. vii, 413 p. Appendixes.

Seminar was held in October 1973 at Seattle, Washington. Many contributors; subjects include modeling, faunal survey, sampling theory, pesticides, and oil and vinyl chloride pollution. References given at end of each contributed section.

985 Federal Working Group on Pest Management. GUIDELINES ON SAMPLING AND STATISTICAL METHODOLOGIES FOR AMBIENT PESTICIDE MONITORING. Washington, D.C.: Government Printing Office, 1974. 9 chaps., separately paginated.

Small book with the following chapter titles: "Statistics and Study Design"; "Air"; "Soil"; "The Hydrologic Environment"; "Estuaries"; "Fresh Water Fish"; "Wildlife"; "Food and Feeds"; "Human Tissues."

986 Gill, R., et al. "Regional Monitoring of Heavy Metal Pollution." CHEMOSPHERE 2 (1975): 113-18.

Describes manner of establishing a monitoring system in southwest England for determining airborne metals. Describes as well a technique of using moss (_Sphagnum_) bags as traps.

987 Henderson, C., et al. "Organochlorine Insecticide Residues in Fish--Fall 1969. National Pesticide Monitoring Program." PESTICIDES MONITORING JOURNAL 5 (1971): 1-11.

Residue samples from fifty stations were analyzed for eleven insecticides, lipids, and PCBs. Total DDT ranged up to 57.8 ppm, wet weight. Dieldrin residues were found in most samples; highest value was 1.59 ppm. This type of detail is issued regularly in this journal.

988 Moore, Norman W. "Pesticide Monitoring from the National and International Points of View." AGRICULTURE AND ENVIRONMENT (Amsterdam) 2 (1975): 75-83.

Author is British. This is an excellent general survey of three aspects of monitoring: (1) surveillance for significant changes in populations and ecosystems; (2) measurements of residues to determine effectiveness of control measures; (3) emphasis on wild species which for biological reasons are good pollution indicators.

989 Thomas, William A., ed. INDICATORS OF ENVIRONMENTAL QUALITY. New York: Plenum Press, 1972. x, 275 p.

Proceedings of a symposium during the American Association for the Advancement of Science meeting in Philadelphia, Pennsylvania, 26-31 December 1971. Twenty contributions cover major types of environmental pollution. Of importance are the "indicator" papers on the following categories: humans, John M. Finklea et al.; aquatic communities, Ruth Patrick; plants, C. Stafford Brandt; soil quality, Richard H. Rust et al.; and plant ecology, David F. Grigal.

Section 15

ANIMAL RESISTANCE TO TOXICANTS

Both target and nontarget species in a chemically saturated area may quickly evolve populations which will not respond normally to chemical treatments. The "solution" is to increase dosage, to withdraw the chemical, or to substitute a new form of control chemical. The first choice hastens increased resistance; the second increases environmental hazard and, very likely, leads to regulatory violations; and the third, we have learned by experience, is only a palliative, for either "cross-resistance" to a related chemical to which the organism has not been exposed quickly occurs, or resistance ultimately develops to the substituted chemical. The ideal solution lies with chemicals formed by the animals themselves, in disease reactions or predatory-prey interactions. The term biological control describes practical measures in which one organism is consciously pitted against a pest species. Integrated pest control employs not only the methods of biological control but also those of crop rotation, different harvesting methods, and appropriate timing of chemical treatment.

990 Boyd, C.E., and Ferguson, Denzel E. "Spectrum of Cross-Resistance to Insecticides in the Mosquito Fish, *Gambusia affinis*." MOSQUITO NEWS 24 (1964): 19-22.

Fish known to be resistant to DDT, BHC, DDD, and endrin were exposed to thanite, dilan, methoxychlor, kelthane, perthane, chlordane, and strobane. There were differences in susceptibility but fish showed marked cross-resistance to strobane and chlordane.

991 ______. "Susceptibility and Resistance of Mosquito Fish to Several Insecticides." JOURNAL OF ECONOMIC ENTOMOLOGY 57 (1964): 430-31.

Gambusia (mosquito fish) showed both resistance and cross-resistance to DDD, endrin, aldrin, dieldrin, heptachlor, toxaphene, and lindane. Resistance in fish populations previously exposed to insecticides is genetically based.

992 Chambers, J.E., and Yarbrough, J[ames]. D. "Parathion and Methyl Parathion Toxicity to Insecticide-resistant and Susceptible Mosquito Fish (Gambusia affinis)." BULLETIN OF ENVIRONMENTAL CONTAMINATION AND TOXICOLOGY 11 (1974): 315-20.

Fish populations resistant to methyl parathion tolerated 1.3-fold more of this chemical than did nonresistant populations. Fish were about 1.6-fold more tolerant to parathion in spring, but not in fall. Difference is related to rapid dissipation of this class of compounds. Contrast with chlorinated hydrocarbons.

993 Fabacher, D.L., and Chambers, H. "Rotenone Tolerance in Mosquito Fish." ENVIRONMENTAL POLLUTION 3 (1972): 139-41.

Rotenone resistance is very rare. Described is a case in vertebrates wherein resistant fish are 1.8-fold more tolerant to same concentration as susceptible strains.

994 Ferguson, Denzel, et al. "Comparative DDT-Tolerance of Cotton Rats from Three Areas Differing in Past Exposure to Pesticides." JOURNAL OF THE MISSISSIPPI ACADEMY OF SCIENCES 11 (1965): 229-33.

Mortality differences in cotton rats from treated and untreated areas were not detectable in feeding experiments.

995 ______. "Resistance to Chlorinated Hydrocarbon Insecticides in Three Species of Freshwater Fish." BIOSCIENCE 14 (1964): 43-44.

Shiners, bluegill, and mosquito fish in waters adjacent to treated field are resistant to toxaphene, aldrin, dieldrin, and endrin, but not DDT. Striking resistance to endrin, ordinarily highly toxic to fish.

996 ______. "Tolerances of Two Populations of Fresh Water Shrimp to Five Chlorinated Hydrocarbon Insecticides." JOURNAL OF THE MISSISSIPPI ACADEMY OF SCIENCES 11 (1965): 235-37.

Compounds tested were DDT, toxaphene, aldrin, dieldrin, and endrin. Shrimp in bayous draining treated fields were one to three times more tolerant of toxicants than those from untreated areas.

997 Georghiou, G.P. "The Evolution of Resistance to Pesticides." ANNUAL REVIEW OF ECOLOGY AND SYSTEMATICS 3 (1972): 133-68.

Excellent review; two hundred and nine references. Broadly based and biological in its emphasis, with practical applications. Graphic illustrations including new material.

998 Finley, M.T., et al. "Possible Selective Mechanisms in the Development of Insecticide-resistant Fish." PESTICIDES MONITORING JOURNAL 3 (1970): 212-18.

Selection pressure from residues present in food chains minimal compared with direct exposure. DDT and toxaphene most important selectively.

999 Perry, A.S. "Insecticide Resistance in Insects and Its Ecological and Economic Thrust." In SURVIVAL IN TOXIC ENVIRONMENTS, edited by M.A.Q. Khan and J.B. Bederka, Jr., pp. 399-445. New York: Academic Press, 1974.

Lengthy, detailed review; eleven pages of references. Considers social aspects of use of broad-spectrum insecticides, notably in disease control in tropical latitudes. Recommends integrated insect control system along with social changes. Excellent review.

1000 Rosato, Peter, and Ferguson, Denzel E. "The Toxicity of Endrin-resistant Mosquitofish to Eleven Species of Vertebrates." BIOSCIENCE 18 (1968): 783-84.

This little paper is a classic. Fishes, frogs, turtles, and birds which were force-fed endrin-resistant gambusia all died. There are frightening ecological predictions based on the resistance phenomenon.

1001 Vinson, S.B., and Plapp, F.W., Jr. "Third Generation Pesticides: The Potential for the Development of Resistance by Insects." JOURNAL OF AGRICULTURAL AND FOOD CHEMISTRY 22 (1974): 356-60.

Certain strains of insects that show resistance to insecticides also show cross-resistance to insect juvenile hormone mimics or analogs. Present experience suggests that through selective pressures from the use of third generation insecticides (third substitution of synthetic chemical exposure), high levels of resistance may develop. This article is part of a symposium published in same issue.

1002 Vinson, S.B., et al. "Resistance to DDT in the Mosquito Fish, _Gambusia affinis_." SCIENCE 139 (1963): 217-18.

First demonstration of insecticide resistance in fishes under field conditions (waters adjacent to cotton fields).

1003 Wells, M.R., and Yarbrough, James D. "In Vivo and in Vitro Retention of [^{14}C] Aldrin and [^{14}C] Dieldrin in Cellular Fractions from Brain and Liver Tissues of Insecticide-resistant and Susceptible Gambusia." TOXICOLOGY AND APPLIED PHARMACOLOGY 24 (1973): 190-96.

Illustrates manner of explaining mechanisms of resistance in fish. Study suggests that there is a membrane barrier to both aldrin and dieldrin in brains of resistant fish.

1004 Woodham, D.W., and Bartlett, F.J. "White-fringed Beetles: Chemical vs. Biological Tests to Determine Resistance to Dieldrin in Larvae." JOURNAL OF ECONOMIC ENTOMOLOGY 66 (1973): 283-84.

Authors show methods of assessing insecticide resistance. In this study, chemical analysis allowed quick determination; biological tests required 120 days.

1005 Yarbrough, James D. "Insecticide Resistance in Vertebrates." In SURVIVAL IN TOXIC ENVIRONMENTS, edited by M.A.Q. Khan and J.B. Bederka, Jr., pp. 373-97. New York: Academic Press, 1974.

Technical review centered on distinguishing differences in cell membranes and brain barriers in resistant and nonresistant fishes.

1006 Young, R.G. "Enzyme Action on Halogen-containing Pesticides." In CHEMICAL FALLOUT, edited by Morton W. Miller and George G. Berg, pp. 315-24. Springfield, Ill.: Charles C. Thomas, Publisher, 1969.

Degradation of DDT in animal tissue is much lower in nonresistant forms although it normally occurs.

Section 16

ALTERNATIVE METHODS OF BIOTIC CONTROL

This subject occupies the attention of thousands of scientists and regulatory officials. It is desirable, of course, to minimize or eliminate the use of chemicals in biological population regulation. The prospects for doing so totally are dim. The prospects for achieving a great deal of success in this special kind of biological engineering are excellent, however. Efforts to integrate the best elements of chemistry, biology, and production practices are now widely appreciated. Greater knowledge of techniques and better cooperative advisory systems will ensure the best possible achievement with these alternate methods of control.

1007 Andres, L.A., and Bennett, F.D. "Biological Control of Aquatic Weeds." ANNUAL REVIEW OF ENTOMOLOGY 20 (1975): 31-46.

> Principles are nearly the same for aquatic as for terrestrial weeds, although seemingly less stringent host specificity is required. More needs to be known about pathogens. Three successful example are cited.

1008 Borkovec, A.B. "New Aspects of Organophosphorus Pesticides. VI. Organophosphorus Insect Chemosterilants." RESIDUE REVIEWS 53 (1974): 67-77.

> Chemical sterilization is much studied as an alternative to broadcast, nonspecific toxins. This is a basic paper on the subject. Alkylating derivatives of phosphoric and phosphorothioic acids and cyclic azaphosphorines are chemical cytostats and mutagenic agents selectively active on insect gonadal tissue. Chemical structures compared with biological effects. No applications as yet.

1009 Conway, Gordon R. "Better Methods of Pest Control." In ENVIRONMENT--RESOURCES, POLLUTIONS, AND SOCIETY, edited by W.W. Murdoch, pp. 355-78. 2d ed. Sunderland, Mass.: Sinauer Associates, 1975.

Problems in use of pesticides in insect control and alternative methods that are or could be used. Good review and referencing.

1010 David, W.A.L. "The Status of Viruses Pathogenic for Insects and Mites." ANNUAL REVIEW OF ENTOMOLOGY 20 (1975): 97-118.

Recent studies relevant to uses in insect control are reviewed. Contains 231 references.

1011 Davidson, G. GENETIC CONTROL OF PESTS. New York: Academic Press, 1974. x, 158 p.

Emphasizes that much work has been done to replace nonspecific, synthetic chemical means of pest control. Summarized herein are methods which entail mass rearing of particular species, irradiation, chemosterilization or hybridization, and, finally, release into the wild population. Genetic manipulation is also discussed. Such methods require either regional coordination among private interests or large-scale governmental programs.

1012 DeBach, Paul. BIOLOGICAL CONTROL BY NATURAL ENEMIES. London and New York: Cambridge University Press, 1974. 334 p.

Excellent review of chemical misuse in pest insect control and the biological methods that can be employed for the same purposes. Historical review of biological control and a statement of its present relationship to integrated control practices. Final chapter entitled "Escape from the Pesticide Dilemma."

1013 DeWilde, J. "Insect Population Management and Integrated Pest Control." AMBIO 4 (1975): 105-11.

General article describing advantages of an integrated control system based on ecological relationships in crop systems.

1014 Jacobsen, Martin, ed. "Alternate Methods of Insect Control to Improve Environmental Quality." ENVIRONMENTAL LETTERS 8 (1975): 1-88.

Five papers by authorities in their subject (e.g., Robert van den Bosch on biological control). Contents include appraisals of integrated control, use of insect diseases, sex pheromones, sex sterilization, and juvenile hormones. Extensive referencing.

1015 Lofgren, C.S., et al. "Biology and Control of Imported Fire Ants." ANNUAL REVIEW OF ENTOMOLOGY 20 (1975): 1-30.

Important technical and historical review of a major control effort. Broad-spectrum pesticide controls failed; biological controls are now sought. One hundred and twenty-seven citations.

1016 MacClellan, C.R. "Codling Moth Populations under Natural, Integrated, and Chemical Control on Apples in Nova Scotia (LEPIDOPTERA: OLETHRENTIDAE)." CANADIAN ENTOMOLOGIST 104 (1972): 1397-1404.

Young commercial orchards require chemical treatment but older require chemicals only occasionally. Predators few in chemically treated orchards. This is a sample of the work at Kentville Research Station, Nova Scotia, whose purpose has been to reduce chemical use.

1017 MacPhee, A.W., and MacClellan, C.R. "Ecology of Apple Orchard Fauna and of Integrated Pest Control in Nova Scotia." In PROCEEDINGS TALL TIMBERS CONFERENCE ON ECOLOGICAL ANIMAL CONTROL BY HABITAT MANAGEMENT, 25-27 February 1971, pp. 197-208.

An example of alternative methods of insect control. The Kentville, Nova Scotia, station of the Canadian Department of Agriculture has endeavored to reduce chemical use in insect control, thereby reducing costs of production, environmental hazard, and pesticidal side effects in faunal management. See references in this paper for methods of control.

1018 Ruesink, W.G. "Status of the Systems Approach to Pest Management." ANNUAL REVIEW OF ENTOMOLOGY 21 (1976): 27-44.

Ruesink developed models in existing pest management systems. Major review with 173 references.

1019 Stairs, G.R. "Pathogenic Microorganisms in the Regulation of Forest Insect Populations." ANNUAL REVIEW OF ENTOMOLOGY 17 (1972): 355-72.

Review of various groups of microbial pathogens and control status relative to insect numbers.

1020 Stern, Vernon M., et al. "The Integrated Control Concept." HILGARDIA 29 (1959): 81-101.

The classic paper in which integrated pest control ideas were formalized. Requires using chemical, biological, and cultural methods in supervised manner.

1021 Wade, N. "Insect Viruses: A New Class of Pesticides." SCIENCE 181 (1973): 925-28.

The polyhedrosis virus of _Heliothis zea_, a noctuid moth known also as cotton bollworm, has been certified for use by the Environmental Protection Agency. The difficulty with viruses is their continued specificity. It is not known if they will mutate or if they can control pest populations indefinitely without hazard.

Perhaps with this EPA registration we see the beginnings of a new group of control agents supplanting chemical pesticides. Hopefully, there will not be too many problems with this approach to pest control.

1022 Wapshore, A.J. "A Protocol for Programmes for Biological Control of Weeds." PANS (London) 21 (1975): 295-303.

A London-based journal with involvements in tropical productivity. Wapshore describes logical procedures for biological control of weeds and examples of failure where protocol was not followed.

1023 Weatherston, J., and Rethakaran, A. "The Potential of Autocides and Microorganisms as Ecologically Acceptable Agents for the Regulation of Spruce Budworm Infestations." JOURNAL OF ENVIRONMENTAL QUALITY 4 (1975): 294-303.

Major review (fifty-one references) of new approaches to the control of the spruce budworm, the most serious pest in North American forests. Autocides are pheromones and hormones. Microorganisms under study are viruses, fungi, bacteria, and microsporidia. Authors conclude that to date nothing totally replaces toxic chemicals, but that integration of these living materials with conventional chemical control practices should be followed.

APPENDIXES

Appendix 1

ABBREVIATIONS AND TERMS

ACh	acetylcholine; basic alkaloid used in nervous transmission
Accumulation	storage and concentration of a chemical in tissue to an amount higher than intake of the chemical
Acute	immediate, usually severe effect; usually limited to effects of a few days duration following a single dosage or other exposure to a chemical
Ad libitum	free acceptance of food materials; as much as wished
Carcinogen	any cancer-producing substance
Carcinogenicity	quality of being cancer producing
Che	cholinesterase; enzyme normally present to destroy ACh, so that nervous impulses may be transmitted
Cholinesterase inhibitor	any chemical allowing buildup of acetylcholine, thereby interfering with transmission of nerve impulses
Chronic	long-range effects which may or may not be severe following multiple dosage or exposure to a chemical; usually implies exposure of several weeks to six months or longer
Cumulative action	increasingly severe effects due either to storage and concentration of a chemical in a particular organ, or to a physiological condition sustained below its threshold of proper performance
CNS	central nervous system

Abbreviations and Terms

Food chain	the series of food sources leading back to the original energy producer
g	gram
EC	effective concentration; produces some reaction
EC50	effective concentration affecting 50 percent of a population
Epoxidize	to convert to a compound in which an oxygen atom is joined to two carbon atoms in a chain to form a bridge
ha	hectare; about 2.4 acres
Half-life	in pollution studies the time required for half the original chemical to dissipate from a given site or to lose its chemical identity. Term is borrowed from decay rates of radioactive compounds
Indirect effects	environmental changes resulting from chemical usage which in turn influence dependent animals
IP	intraperitoneal; manner of injection; into abdominal cavity
IV	intravenous; manner of injection; into veins
kg	kilogram; one thousand grams
l	liter
lbs/ac	pounds per acre
LC	lethal concentration; applies to water or gasses
LC50	concentration lethal to 50 percent of a test population
LD	lethal dose; applied to animals
LD50	doses lethal to 50 percent of a test population of animals
m	meter
mg/kg	milligrams (of chemical) per kilogram of body weight; relates dosage to a standard weight rather than to varying body weights of individual test animals

ml	milliliter; one thousandth of a liter
microg	microgram; one millionth of a gram
Mutagen	any substance which alters gene character, producing mutation
Mutagenicity	quality of altering gene character
ppb	parts per billion
ppm	parts per million; applied to concentrations in gas, vapor, or liquids; to foods; or to amounts in tissue
ppt	parts per trillion
Primary poisoning	effect produced by direct exposure to a chemical in its original formulation
Secondary poisoning	effect produced in a second animal by ingesting parts or all of a poisoned animal; usually depends on a chemical that is stable
Subacute	effects within a few days to a month; often implies less severe effects than "acute"; usage not precise
Subchronic	effects from short-term multiple dosage or exposure; usually means exposure for less than three months
Teratogen	any substance producing embryonic deviations or malformations
Teratogenicity	quality of producing embryonic deviations
Threshold	the chemical amount or the physiological state that must be reached before a given reaction sets in
TL_m	median lethal threshold; usually applied to fishes
96-hr-TL_m	TL_m after 96 hours; standard measurement of toxicity to fish; adds time factor in reaction. Any period of time could be used
Translocation	movement of chemical within a plant or animal; usually refers to systemic herbicides and insecticides that are moved from the point of contact on a plant to other regions of the plant
Trophic concentration	transfer of stored residues along a food chain; sequence of producing and consuming energy layers in an ecosystem

Appendix 2

COMMON AND CHEMICAL NAMES OF COMPOUNDS MENTIONED MOST FREQUENTLY IN TEXT

Aldrin	1,2,3,4,10,10-hexachloro-1,4,4a,5,8,8a-hexahydro-exo-1,4-endo-5,8-dimethanonaphthalene
Arochlor 1260	PCB, approximately 60 percent chlorine
Azinphosmethyl (Guthion)	0,0-dimethyl S [4-oxo-1,2,3-benzotriazin-3 (4H) ylmethyl] phosphorodithioate
BHC (Benzene hexachloride)	1,2,3,4,5,6-hexachlorocyclohexane (mixture of isomers). Commercial product contains several isomers of which "gamma" is most active as an insecticide
Carbaryl (Sevin)	1-naphthyl methylcarbamate
Chlordane	1,2,3,4,5,6,7,8,8-Octachlor-2,3,3a,4,7,7a-hexahydro-4,7-methanoindane. The technical (commercial) product is a mixture of several compounds including heptachlor, chlordene, and two isomer forms of chlordane
Chlorpyrifos	see Dursban
Dalapon	2,2-dichloropropionic acid, sodium salt
2,4-D	2,4-dichlorophenoxyacetic acid
2,4,5-T	2,4,5-trichlorophenoxyacetic acid
p,p'-DDD (TDE)	2,2-bis(p-chlorophenyl)-1,1-dichloroethane
DDE	dichlorodiphenyl dichloro-ethylene (degradation product of DDT). Main component (p,p'-DDE): 1,1-Dichloro-2,2 bis(p-chlorophenyl) ethylene o, p'-DDE: 1,1-Dichloro-2(o-chlorophenyl)-2-(p-chlorophenyl) ethylene

DDT	main component (p,p'-DDT): 1,1,1-Trichloro-2, 2-bis(p-chlorophenyl) ethane. Other isomers are possible and some are present in the commercial product. o,p'-DDT: [1,1,1-Trichloro-2-(o-chlorophenyl)-2-(p-chlorophenyl) ethane]
DDVP	see Dichlorvos
Dichlorvos	2,2-dichlorovinyl dimethyl phosphate
Diazinon	0,0-diethyl 0-(2-isopropyl-6-methyl-4-pyrimidinyl) phosphorothioate
Dieldrin	not less than 85 percent of 1,2,3,4,10,10-hexachloro-6,7-epoxy-1,4,4a,5,6,7,8,8a-octahydro-1, 4-endo-exo-5,8-dimethanonaphthalene
Dursban	active ingredient, chlorpyrifos-0,0-diethyl 0-(3,5, 6-trichloro-2-pyridyl) phosphorothioate
Endrin	1,2,3,4,10,10-hexachloro-6-7-epoxy-1,4,4a,5,6, 7,8,8a-octahydro-exo-1,4-exo-5,8-dimethanonapthalene
Fenthion	0,0-dimethyl 0-(3-methyl-4-methylthiophenyl) phosphorothioate
Guthion	see Azinphosmethyl
Heptachlor	1,4,5,6,7,8,8-heptachloro-3a,4,7,7a-tetrahydro-4,7-methanoindene
Heptachlor epoxide	1,4,5,6,7,8,8-heptachloro-2,3-epoxy-3a,4,7,7a-tetrahydro-4,7-methanoindane
Kepone	1,3,4-Metheno-2H-cyclobuta [cd] pentalen-2-one, decachlorooctahydro-
Lindane	gamma isomer of benzene hexachloride (1,2,3,4,5, 6-hexachlorocyclohexane) of 99+ percent purity
Malathion	S-[1,2-di(ethoxycarbonyl)ethyl] 0,0-dimethyl phosphorodithioate
MCPA	4-chloro-2-methylphenoxyacetic acid
Methoxychlor	1,1,1-trichloro-2,2-bis(p-methoxyphenyl) ethane

Methyl parathion	phosphorothioic acid, 0,0-dimethyl 0-p-nitrophenyl ester
Mirex	dodecachlorooctahydro-1,3,4-metheno-1H-cyclo-buta [cd]pentalene
Paraquat	bipyridinium compounds, 1,1-dimethyl-4,4'___ . . . dimethyl sulfate or dichloride
Parathion	0,0-diethyl 0-p-nitrophenyl phosphorothioate
Parathion-methyl	0,0-dimethyl 0-p-nitrophenyl phosphorothioate
Phosphamidon	2-chloro-2-diethylcarbamoyl-1-methylvinyl dimethyl phosphate
Polychlorinated biphenyls (PCB)	biphenyl polychlorinates (approximately 20 to 60 percent chlorine, various members)
Polyvinylchloride (PVC)	polymerized substituted ethylene
Sevin (carbaryl)	1-naphthyl N-methylcarbamate
Simazine	s-Triazine, 2-chloro-4,5-bis(ethylamino)-
TDE	see DDD
Toxaphene	chlorinated camphene containing 67 to 69 percent chlorine

INDEXES

AUTHOR INDEX

Many technical papers and books have multiple authors. This index lists authors, editors, and compilers, singly whether first or second in order. Where appropriate, the name of a governmental agency or other sponsoring source or editors are also listed. All authors in text are referenced by entry number only; no page numbers are referenced. Alphabetization is letter by letter.

A

B

C

D

E

F

G

H

I

J

K

N

O

P

R

S

W

Y

Z

SUBJECT INDEX

All references are to entry numbers in text. Since most entries contain many technical details, many text references are included under multiple categories. Underlined entries refer to main sections identified in the table of contents. Alphabetization is letter by letter.

A

Subject Index

C

D

E

F

G

H

I

N

O

Q

R

S